A CENTURY OF SCIENCE IN THE UNIVERSITY OF GLASGOW

Edited
by
Robert Young Thomson

UNIVERSITY
of
GLASGOW

Design and production: Lynn Bell
Cover design: Graeme Campbell, Hunterian Museum
Printed by Bell & Bain Ltd, Glasgow
ISBN: 0 85261 412 8

University of Glasgow, Glasgow G12 8QQ

Contents

Acknowledgements
Information & Publicity Office; Graeme Campbell and Ron Harrison, Hunterian Museum; Trevor Graham and the University Photographic Unit; University Archives; Professor R. Whitehead, Department of Physics & Astronomy; Ian Ramsden, Medical Illustration Unit.

Preface

This collection of essays is published in celebration of the Centenary of the establishment of the Faculty of Science in the University of Glasgow. Faculties of themselves are bloodless legislative contrivances, technically, at Glasgow, mere sub-committees of the Senatus Academicus. Yet the establishment of the new Faculty at Glasgow and simultaneously two of her sister universities did mark an intellectual initiative, a realisation that a new world had opened beyond the classics and mathematics, philosophy and logic, divinity and law which had since the middle ages provided the basis of learning. The Faculty has had a varied history. It originally embraced engineering, which very wisely hived off in 1923, and at one time it offered degrees in applied science, notably pharmacy and agriculture. It has, especially over the past 30 years, greatly diversified. Yet intellectually one can still trace the inheritance of the eight professors who were appointed to constitute it on 5 June 1893. The present volume is in no sense an attempt to offer a comprehensive account of all that Faculty has attempted and achieved over the past hundred years. An account of the legislation it has enacted would be unprofitable to record and exceedingly dull to read. Instead we have invited eleven contributors to write accounts of the manner in which the original eight departments, which still form the core of the Faculty, have developed. They have been invited to use their own discretion as to how they did so. Intellectual history is not to be constrained by the principles of Procrustes. But it is hoped that their essays will provide a record of permanent value of a period of important development in the University's history. Several of the authors have been *cives Universitatis Glasguensis* for almost half of Faculty's existence and were pupils of men and women whose recollections would have extended much further back.

There is a certain historical parallel between 1893 and 1993. Then, even more than now, the ancient Scottish universities were under critical scrutiny. Those of us who are apprehensive of our future now may take heart from what has been achieved over the past century.

The History of the Faculty

Robert Young Thomson

The Faculty of Science of the University of Glasgow was established on 5 June, 1893. Though the idea of such a development had been canvassed earlier, most notably in the evidence offered to a Commission of Enquiry in 1877, opinion within the University had been divided. Of the professors whose opinions had been canvassed, McKendrick (Physiology), Blackburn (Mathematics), Young (Zoology) and Allen Thomson (Anatomy) had been in favour, but Grant (Astronomy) and, far more important, William Thomson, the future Lord Kelvin, had been against. So the Faculty was established, not by the University's own volition, but by Ordinance of the Commissioners appointed under the Universities (Scotland) Act 1889. Those of us fortunate to have enjoyed the *douceur de vivre* of the '50s, '60s and '70s of the present century, when the autonomy of universities was an article of faith and politicians confined themselves to distributing largesse and establishing new foundations in green fields, may be startled to discover that in the last century Parliament, which contributed little to their support, had not the least hesitation in intervening in the minutiae of life in the four ancient Scottish universities. And so it was that the Faculty of Science was established by the Commissioners under the Act. Supervision of the universities first by church, then by state, had ample precedent. Glasgow was visited repeatedly in the latter half of the 17th and the early years of the 18th century. But it was left to the Victorians to attempt a root and branch reform. The Universities of Scotland Acts of 1858 and 1889, though liberalised by a third Act in 1966, effectively established the constitutional arrangements by which the four ancient Scottish universities are still governed. Though for our purposes the 1889 Act is more important, in that it led to the establishment of the Faculty, its form followed a pattern set by the 1858 Act. The substance of both was concerned with the mechanism of government, but both set up Commissions which were endowed with far-reaching powers to prescribe how the universities should discharge their teaching functions. The names of the Commissioners were set out in each Act and the duration of their authority prescribed; five years under the Act of 1856, three, with the possibility of extension, under that of 1889. The Commissioners of 1858 made only limited use of their powers. The Ordinances promulgated under their authority form a modest volume of just over 200 pages, largely taken up with regulations governing various charitable foundations. In academic terms, their major achievement was to lay down regulations for degrees in Arts and Law which were to be common to all four universities, and for degrees in Medicine which were intended to meet the special circumstances of each of the four. The Commissioners under the 1889 Act are of more immediate significance to our present purpose. Of the 16 persons named, only three will be familiar to the modern reader. Sir William Thomson, the future Lord Kelvin, is the intellectual giant among them. John Patrick Crichton-Stuart, third Marquis of Bute, the eccentric convert from the Church of Scotland to Roman Catholicism, has a lasting memorial to his munificence in the splendour of Bute Hall. Alexander Bruce, Lord Elgin, was a future Viceroy of India (though not perhaps the most forceful holder of that office). But the remainder included some whose merits are recorded in the *Dictionary of National Biography.* Alexander Smith Kinnear, who was chairman, was one of the most distinguished appellate judges to have graced the Scottish bench. Sir Patrick Heron Watson was an eminent Edinburgh surgeon. Sir Arthur Mitchell was not merely a Scottish Commissioner in Lunacy but a distinguished antiquary and the first Rhind Lecturer in Archaeology at Edinburgh. Samuel Herbert Butcher was Professor of Greek at Edinburgh and co-author with Andrew Lang of a much-admired translation of the *Odyssey.* Alexander Craig Sellar, the son of that Patrick Sellar who achieved notoriety in the highland clearances, was Professor of Latin at Edinburgh. Francis Richard Sandford was a distinguished civil servant. And Sir Harry Enfield Roscoe, an eminent chemist, had a prominent hand in establishing what we would now call the 'redbrick' universities in the north of England.

The powers entrusted to the 1889 Commissioners broadly followed those granted to their predecessors 30 years before, but they were exercised much more widely. The 1858 Commissioners had contented themselves with promulgating only nine general Ordinances (ie, applying to all four universities) in their five years tenure of office; their successors issued no less than 41 over a period of eight years. The seventh of these, dated 3 February 1892, laid down regulations for degrees in sciences. A separate Ordinance dated 5 June 1893 instituted faculties of science at Glasgow, Aberdeen and Edinburgh (but not at St Andrews) and prescribed the professors who should constitute them. At Glasgow these were the professors of Mathematics; Natural Philosophy; Astronomy; Chemistry; Natural History; Botany; Anatomy; Physiology; Civil Engineering and Mechanics; and Naval Architecture.

The study of science at Glasgow of course far antedates the institution of the Faculty. The original Arts curriculum at Glasgow closely resembled that already taught at St Andrews, which in turn followed the model established at Paris. In origin this was based on the, already ancient, medieval trivium (grammar, logic and rhetoric) and quadrivium (mathematics, geometry, astronomy and music). In practice, few medieval teachers could go beyond the elementary aspects of the quadrivium and, within the trivium, logic, especially at Paris, had gained pride of place. Logic was ,in the middle ages, grounded on the works of Aristotle. The logic of Aristotle led the student to the study of philosophy as expounded in a number of works erroneously attributed to him. This was very much what in traditional Glasgow parlance would be called natural as opposed to moral philosophy. The original method of teaching was that each of the three regents in arts took one class - all the students who entered in a given year-through the entire three-year curriculum, and then started all over again with a fresh entry. This system was abandoned in the Nova Erectio of 1577. Thereafter the three regents became specialists, the first in rhetoric and Greek grammar, the second in dialectic, logic and moral philosophy, and the third in what was then called 'physiology' but which really embraced all the little that was known of the natural sciences. The regents were thus embarked on the course which would culminate in their transformation into professors. That transformation was completed in 1727 when the regency system was finally given up. At that date the University had grown to the point where 'philosophy', in its broadest sense, was taught by three regents. They were invited to choose the subjects they were in future to profess. Gershom Carmichael, who in his own day enjoyed a considerable reputation, elected to teach moral philosophy; John Loudon, a lesser figure, opted for logic. Natural philosophy, what we would now call physics, fell to the third regent, Robert Dick, because, it has been said, he was the only one of the three who knew sufficient mathematics. Natural philosophy, as that term was understood in the 18th century, was narrower in scope than had been the 'physiology' of the middle ages. Galileo, Kepler and Newton, among others, had established the pre-eminence of physics and astronomy among the sciences. They alone could be investigated and their phenomena described in mathematical terms. The sciences which could not as yet be treated mathematically, in particular chemistry and biology, lay in the province of the medical man rather than the scholar. There were obvious reasons for this. Once the search for the philosopher's stone had been given up, the link between chemistry and materia medica became even more important; and one of the main reasons for the study of botany was the number of drugs derived from the physic garden.

The Faculty of Science therefore derived from the marriage of two traditions: the physical science taught in the Faculty of Arts and the chemistry and biology taught in the Faculty of Medicine. Curiously, the B.Sc. degree antedated the new Faculty. It had been instituted by the University in 1872 but it was not necessarily a degree in science in the modern sense, viz. physical or natural science, but rather in the older, and perhaps even then archaic, sense of knowledge gained by study. Originally five alternative courses of study were offered. These were in: law; biological sciences; geological science; engineering science; and arts.

To the modern reader, the new degree looks very much like a means of escape from the rigid curriculum for the M.A. laid down by the 1858 Commissioners. This allowed the student absolutely no choice: he must take not less than two sessions of humanity (ie, Latin), Greek and mathematics, one each of logic, moral philosophy and natural philosophy, plus a course in English literature. The B.Sc. offered a much greater range of courses, including chemistry, zoology, anatomy, geology, anatomy and physiology, and far greater freedom of choice.

The Commissioners of 1889 revised the regulations, both for the M.A., and the B.Sc. Partly for constitutional reasons, but partly also because their work commended itself to subsequent generations, their thinking determined the form of Scottish Arts and Science degrees for many years, and indeed its influence is not wholly exhausted even now.

Before passing on to the B.Sc. regulations it is well worth looking at those for the M.A. which were promulgated exactly a year earlier. The rigid curriculum of the 1858 Commissioners was abandoned for a flexible arrangement, but one whose variations were constrained by requirements which were both ingenious and wise. By a practice in which may be seen the shadow of trivium and quadrivium, the curriculum had to comprise seven courses. The Scottish tradition demanded a broad education. Accordingly, candidates were required to study Latin or Greek; English or a modern language or history; logic or moral philosophy; and mathematics or natural philosophy. But their studies should centre on a theme: they were therefore required to study both Latin and Greek; or both logic and moral philosophy; or two out of mathematics, natural philosophy and chemistry. The regulations for the B.Sc. show a similar concern for, on the one hand, flexibility of choice, and, on the other, intellectual coherence. The B.Sc. was no longer to be a versatile, and perhaps vocational, alternative to the rigid M.A. It was to be a degree in its own right, testifying to a competence in the physical and natural sciences. As in the M.A., the candidate had to complete the traditional seven courses. He was required to pass a first science examination in natural philosophy, chemistry and either mathematics or biology. He might then proceed to attempt on the higher standard (ie, that required in the Honours M.A.) three out of: mathematics; natural philosophy; astronomy; chemistry; anatomy; physiology; geology; zoology; botany. The new degree therefore was equivalent to the Honours M.A.

Both the 1858 Act and its successor of 1889 imposed on the universities a uniformity of practice which persists to this day. The 1858 Act clearly envisages them federating to form a national university for Scotland. And the 1889 Act imposed on them a constitutional mechanism which made any major change in one of them dependent on its submission to all the others and ultimately on the approval of the Queen in Council and its being laid before both Houses of Parliament. These requirements might prove in practice to be formal, but they hardly encouraged enterprise or innovation. It is therefore not surprising that the B.Sc. regulations ordained by the Commissioners in 1893 should have survived unchanged in Glasgow until 1921. In that year a fundamental reform was undertaken by which the degree was split into an Ordinary and an Honours option. The Ordinary was to extend over not less than three sessions and was to comprise five subjects, two of which must be studied over not less than two sessions (the double courses, as they used to be called). The student's choices were subject to constraints; he was obliged to take natural philosophy and chemistry and either zoology or botany; and he was forbidden to select as his two double subjects combinations (such as astronomy and zoology or physiology and geology) which obviously showed no intellectual coherence. He was also required, before entering second year, to submit to Senate for approval a statement giving the subjects he proposed to include in his curriculum. To assist him in this he was provided with a list of no less than forty approved curricula from which to choose. The candidate for Honours had a slightly greater measure of freedom. He need take only four subjects rather than five; he had a distinctly wider choice of subjects; his choice was less constrained by express regulations. But he too was required to have his proposed curriculum approved by Senate before he entered second year and he was, to say the least of it, strongly encouraged to opt for one of the model curricula published for his benefit in the University *Calendar*. He differed from the candidate for the Ordinary degree chiefly in the fact that, instead of taking two double-course subjects for two sessions, he pursued a single principal subject for three or, more usually, four.

These regulations, like their predecessors, proved long-lived, perhaps deservedly so. They corresponded well with the divisions in the sciences as these were widely recognised and as they were represented in Glasgow departments. The range of subjects offered in, say, 1945 had barely altered since 1920 or 1893 or even 1870. To the original list of mathematics, botany, natural philosophy, zoology and geology (until 1903 called natural history), chemistry, anatomy, astronomy, and physiology, there had been the addition only of geography and pathology with bacteriology. This list and the regulations based upon it reflected accurately the academic world's perception of science. The division between the physical and natural sciences was almost total. And on either side of the divide, distinctions were equally clear. Though physics and chemistry shared a wide debatable land of physical chemistry and chemical physics, no-one doubted that they were two very different disciplines. Equally, in the biological sciences the differences between plants and animals and micro-organisms were so great that it was natural that they should be regarded as different provinces. This tidy and satisfying way of organising knowledge began to break down in the 1950s and 1960s, especially in the biological sciences. The rot probably started in biochemistry with the gradual realisation that, despite enormous morphological differences, living organisms share, in varying degrees, a remarkably uniform repertory of chemical structures and mechanisms. Thus a human muscle deriving energy from the conversion of glycogen to lactic acid uses very largely the same mechanisms as a yeast cell fermenting the glucose in grape juice to alcohol and carbon dioxide. A little later the invention of the electron microscope enormously extended our understanding of the fine structure of the cell and demonstrated that, for example, the mitochondria of plants and animals conform to the same basic pattern. Just as the electron microscope permitted visualisation of smaller and smaller structures, so X-ray crystallography made possible the exploration of larger and larger molecules. The attention of biologists shifted from the features that distinguish one organism from another to those phenomena that they all exhibit. Biochemistry, genetics, cell biology assumed a new importance. Traditional boundaries between subjects seemed less important. The question, posed in all seriousness by an eminent Glasgow microbiologist in 1948, whether bacteria should be classed as plants, seemed an irrelevance.

Contemporary developments in the physical and mathematical sciences seem (at least to a biochemist) less revolutionary but nonetheless important. The replacement of the thermionic valve by the transistor made possible the modern computer, a device which, *inter alia*, permitted the solution of problems previously unassailable because of the sheer mathematical labour they demanded. The elucidation from crystallographic data of the three-dimensional structure of complex molecules is an obvious example. At the same time the new quantitative biology gave statistics an importance far greater than it had ever before enjoyed.

These changes were sudden: in 1950 the time-honoured division of the sciences would have seemed entirely rational; in 1960 one was conscious that there was too much new wine for the old bottles. Happily, adaptation was swift. It took two forms. One was the creation of new Chairs in Genetics (1955) and Virology (1958), Cell Biology (1965), Microbiology (1964), Statistics (1965) and Computing Science (1966). The other was in the reform of the B.Sc. regulations. In 1964 a new Ordinance passed through all the stages prescribed by the 1889 Act. It effected a considerable liberalisation. The minutiae of its provisions need not concern us but one or two important innovations merit mention. The range of subjects offered to the Ordinary degree candidate was far wider than before: eight mathematical or physical sciences and ten biological. And though he must still include at least one subject from either group he was no longer obliged to take chemistry and natural philosophy with zoology or botany. And where in the past he had been required to take two of his subjects for two sessions, he was now allowed to take one of them for three sessions. The Honours regulations remained broadly unchanged, save for a corresponding increase in the variety of courses on offer.

The 1964 arrangements were within two years overtaken by the Universities (Scotland) Act of 1966. *Inter alia,* this liberated the four ancient universities from the cumbrous Ordinance procedure whereby the enactments of any of the four had to be circulated for comment to the other three before being submitted for approval by the Privy Council. For the future, the Court of each university was allowed to govern its own affairs by resolution. Glasgow took immediate advantage of its new freedom to liberalise further its B.Sc. regulations by Resolution No.14 of 1967. Again the minutiae need not concern us. The most important innovation was that, for the first time since 1893, a student was allowed to include any arts subject in his curriculum. Four years later a further Resolution (No.54 of 1971) was enacted. This was to a large extent an attempt to systematise the regulations to accommodate the steadily growing list of subjects on offer to students, a process which was carried a stage further by yet another Resolution (No.232) in 1983 which also for the first time made provision for joint Honours in two cognate subjects. Taken together, these enactments from 1964 onward marked the progressive abandonment of prescription in favour of flexibility. The Commissioners in 1893, the University Court in 1921, had sought to limit a student's curriculum to a combination of subjects which they regarded as intellectually coherent. In particular they sought to ensure that the Ordinary B.Sc. should represent a knowledge of what were then regarded as the principal divisions of the natural sciences. That was a reasonable ambition down to, say, 1950. Thereafter the diversification of science rendered it an unrealisable aspiration. Faculty and Senate recognised the inevitable and allowed students an unprecedented freedom in their choice of curriculum. The results could seem bizarre. One student graduated with a B.Sc. in which the principal subject was the history of fine art. But on the whole the new liberalism worked well. The graduate in question went on to an M. Litt. and to the curatorship of a museum in the west of England. The new wine might not match the old bottles, but its quality was as high as ever.

Underlying the changes in the curriculum was an intellectual and social change within the University. In the 1890s, as in previous centuries, the professors were the ordinary teachers. They were supported, in some cases, by one, or even two, lecturers, and by a slightly greater number of assistants. The terms of employment of these, to use the older terminology, junior staff were governed by Ordinance 17 of the 1889 Commissioners. Lecturers were appointed by the University Court for a term not exceeding five years (which was renewable) **either** to teach subjects not already taught **or** to teach branches of their subject not 'usually or fully covered by the Professor's lectures' **or** simply 'to increase the teaching power' in their subject. An assistant was appointed for one year (which was renewable) on the recommendation of the professor whom he was to assist. In 1893 the eight professors in the Faculty of Science (excluding the two professors of Engineering) were supported by five lecturers and 12 assistants. Lord Kelvin, certainly in public esteem the most distinguished physicist of his day, had two lecturers and one assistant, though the latter, J. T. Bottomley, was already a Fellow of the Royal Society. The balance of these arrangements changed only slowly. In 1923 the number of professors had increased to 13 but there were now 22 lecturers and 36 assistants. In 1934 there were still only 13 professors while the number of lecturers had increased to 47 with 16 assistants. Professors had clearly made the transition from being teachers aided occasionally by their juniors to being, in the modern sense, heads of departments. Chemistry alone had, along with its two professors, no less than 16 lecturers. Even in those days, when apparatus was commonly simple and cheap, and when technicians were few, a department of this size needed a head with some skill in the arts of management. The point was made with characteristic forcefulness by one of the most vigorous and successful professors recruited to the Faculty in the 1940s. It was, he averred, his principal duty to create a department in which younger people could do good work. This was not a doctrine to which Kelvin would have subscribed, but it was appropriate then and remains appropriate now. (It is pleasing to note that the professor in question was singularly successful in the objective he set himself.)

Those of us who have known the Faculty for almost half a century cannot but take a certain satisfaction in all that has been achieved within our lifetime. The present-day student is offered a menu of courses and honours options of quite bewildering variety. Our strength in research has never been greater. And the pastoral care we afford our students far surpasses that we could offer in the past. This last is largely the achievement of one of Faculty's most distinguished and devoted servants, the late Jim Collins. The fact that students in Arts and Science did not, unlike those in, say, medicine, follow a prescribed course made it desirable that they should be offered some guidance in the subjects appropriate to their academic objectives. To this end, Court had appointed successively G. R. Peddie and John Thomson as Adviser of Studies in both faculties. This remit was excessive for a single official and the advice offered was necessarily limited. No-one who had the benefit of John Thomson's brisk advice or who witnessed the speed with which he could dispatch a waiting-room of new admissions will forget either experience. In fairness let it be said that, faced by sorrow and hardship, as in the case of a Jewish refugee from Warsaw, his kindness, support, and sympathy were exemplary. But he clearly carried a quite excessive burden, and when Sir Charles Wilson arrived as Principal, the duties he had discharged were divided and Jim Collins' activities were concentrated on the Faculty of Science. He was well equipped for his responsibilities. A physicist by training, he had served in the meteorological branch of the Royal Air force and had subsequently become a schoolmaster. His pedagogical background served him well. Faculty has never lacked distinguished scientists, but few of its members have made a systematic study of educational questions. In the educational debates of the 1960s and 1970s Faculty members looked to Jim to offer a lead, and were never disappointed. The regulations under which we work today are to a great degree the reflection of his ideas. But legislation was the lesser of his achievements. His lasting claim to Faculty's gratitude was his replacement of the previous scheme of a single full-time adviser of studies by a network of part-time advisers, each responsible for around 75 students, charged formally with the duty of advising on choices of course, but expected informally to afford help and support in any of the difficulties, academic, social or emotional, that an undergraduate might face.

Collins' initiative had a profound effect on the ethos of the Faculty. Inevitably academics in general, senior academics in particular, view the undergraduate population from the top of a pyramid. They are familiar with the senior Honours class, a little less familiar with the junior Honours class, and they will have a watchful eye on younger students who show signs of academic promise. What Collins did was to bring to the attention of his advisers, who were carefully recruited from the senior ranks of the academic staff, the broad cross-section of the undergraduates in the Faculty. For many of us, whose own undergraduate careers had been smooth and untroubled, it was a revelation to realise how many of our charges had to struggle to complete their degrees and how many, sadly, fell by the wayside. Standards of pastoral care rose dramatically, not least because of Collins' personal example. He was gifted with a warm heart and outgoing personality and extended his friendship readily and widely. It is hardly surprising that he attracted the loyalty of the advisers he recruited and instilled in them the same attitudes of care and concern for students which he exemplified himself. He died, suddenly and tragically in 1982 on the eve of what should have been a well-earned retirement, but the impulse he imparted has still not been exhausted. It is hardly an exaggeration to say that when he came to the Faculty it was essentially a mere legislative device, a sub-committee of the Senatus Academicus. When he died it was a corporate body with its own administration and a healthy *ésprit de corps.*

In retrospect, what can Faculty have been said to have achieved in its century of existence? Some of the answers are obvious and can be quantified. The Faculty is an enormously larger organisation than the minuscule group of eight professors, with their five lecturers and 12 assistants who first constituted it. At the turn of the century seven students graduated in what was then called 'pure science' as opposed to engineering. Today we look for a hundred times that number. And students entering the Faculty face an almost embarrassing range of specialties. The natural sciences, which only 50 years ago were regarded as the poor relations of classics and mathematics, have assumed a new prominence in Scottish education. In that revolution the Faculty has played a conspicuous role. It has provided, with its sister in Strathclyde, a high proportion of the graduates who teach science in Scottish schools and in the colleges (some now universities) which have flourished so notably in the post-war years. Its impact in these terms on the intellectual life of Scotland has been considerable. So also has been its research output. Those of us whose memories extend back to the 1940s remember vividly how the Second World War impoverished the University, especially in the sciences; how long and slow the uphill pull to recover all that had been lost; and how gratifying it was to find, in perhaps the early 1960s, that we had ceased to be intellectually the poor relations of the great American laboratories. For those of us who teach and study in the Faculty the past two or three decades have been an exciting and rewarding time. One hopes that they may have proved equally so for our students. Certainly a great effort has been put into the improvement of teaching. It is no longer supposed that achievement of first class Honours is evidence of ability to teach anyone anything, guided only by the light of one's inborn intellectual superiority. Equally it is accepted that the methods of instruction hallowed by decades of practice are not necessarily those best suited to the present day. And the humility of the younger generation of teachers has evoked a welcome response from those they teach. Relations between teacher and pupil are amicable, indeed cordial, to a far greater degree than they were 40 or 50 years ago. In those days a lecture might well consist of a teacher transcribing his notes on to a blackboard or reading them at dictation speed, in either case to be copied down by the class, without the least communication of intellectual excitement. Equally, a teacher believed by his audience, perhaps quite unfairly, to have nothing of interest to say, was liable to be barracked without mercy. One recalls from the 1940s a harmless, kindly, middle-aged man who taught an unfashionable aspect of his subject, and who was reduced to writing his entire lecture on the blackboard because he could not make himself audible above the hubbub by which he was assailed. In some respects life has improved: this is one of them.

What hopes does one entertain for the future? Firstly that Faculty will retain a leading place in research and education. That is not to deny the claims of our sister University of Strathclyde or of the newly erected foundations of Paisley and Glasgow Caledonian. Our age, our tradition, our strength confer not rights but obligations. We make no claim to privilege or prestige, but we recognise that we are under a particular duty to promote the advancement of science in the west of Scotland. We have much to contribute to our sister institutions and much to gain from them in return. Secondly, one may hope for the extension of the fresh initiatives of the past 30 years. Prominent among these has been the setting up at all levels of courses taught not by a single department but by a consortium of several. Of the 49 Honours options listed in the current *Prospectus,* 24 are of this nature. And lastly one would hope for a continuation of that steady improvement of standards of teaching and pastoral care of students which has marked the past 30 years. Though much has been achieved, much still remains to be done. University education was within living memory reserved for an elite who qualified by their intellectual attainments and interests, and by the ability and willingness of their parents to make the necessary financial sacrifices. Correspondingly, the graduate, having achieved his degree, could look forward to employment which might not be especially remunerative but which in the Scotland of yesterday was prestigious. Schoolteaching, the profession followed by so high a proportion of Glasgow science graduates, is the obvious example. The future clearly will be different. We have, over the past 20 years, seen an immense expansion in what we might term university-type teaching, culminating recently in a doubling of the number of institutions granted by government fiat the right to style themselves universities. In the future a far higher proportion than in the past will be graduates. This is a situation which will require readjustments in our teaching no less drastic than those imposed on us a century ago.

Robert Young Thomson is Grieve Lecturer in Biochemistry at the University of Glasgow.

Mathematics

Robert A. Rankin

Mathematics has formed part of the curriculum of the University of Glasgow since its very early years, so that it seems appropriate to begin by describing briefly how the subject was taught during this period. Initially it consisted of little more than arithmetic and some elementary geometry, and was apparently taught by the regents to students in their second year. This was probably regarded as a chore by the less scientifically minded regents, and in 1664 a Commission of Visitation that met in Glasgow recommended the appointment of "ane professor of mathematicks which this Universitie formerlie had, and by the erectione and foundatione aucht to have". No evidence, however, for the existence of any previous professor has survived, and it was not until 1691 that George Sinclair was appointed as the first holder of the Chair.

He was a St Andrews graduate and had been at various earlier times a regent in Glasgow. Sinclair was an extremely able and rather curious man, and was one of the most distinguished holders of the Chair. He was the author of books on mathematics, astronomy, navigation, coal and witchcraft. A very practical man, he developed methods for draining coal mines, and supervised the first water supply to be brought to the city of Edinburgh. He was one of the first users (and possibly co-inventor) of the diving bell, with which he assisted, in 1655, in the recovery of canon from the Armada ship, the *Florida*, sunk in Tobermory Bay. His book on witchcraft, *Satan's invisible world discovered*, was very popular and was reissued seven times between 1685 and 1871. An unpublished manuscript entitled *Cacus pulled out of his den by the heels*, and written to rebut criticisms made by James Gregory of St Andrews University, is lodged in the University Library.

Of Sinclair's successors in the Chair, among the most distinguished was Robert Simson (1687-1768), whose edition of Euclid's *Elements*, first published in Glasgow in 1756, ran through numerous editions, the last being published in 1933. He was also known for his skilful restoration of incomplete early Greek mathematical texts, and was the first member of Senate to hold the office of Clerk of Senate, in which capacity he served from 1728 to 1761. Simson held his Chair for 50 years and had several distinguished pupils, the most renowned being Colin MacLaurin, who became Professor of Mathematics, first at Marischal

College, Aberdeen and then at Edinburgh. Our student mathematical society is named after him.

Unlike natural philosophy, mathematics was not a required subject for graduation, despite strong representations by Simson's successor John Williamson and later professors of mathematics. This caused difficulties to those students of natural philosophy who did not have an adequate mathematical background. On more than one occasion, agreement appears to have been reached on the proposal to make some mathematics compulsory, only to be deferred to a later occasion. This became less of a problem with the establishment of Honours degrees in 1826.

After this brief description of the teaching and study of mathematics in earlier times, we come to the year 1893, when the Faculty of Science was founded. It should be pointed out, however, that the degree of B.Sc. had been instituted earlier in 1872, and that the only mathematics classes available to students in the new Faculty were identical with the classes already existing in the Faculty of Arts. This remained the position for nearly 30 years, and even when Honours degrees were eventually introduced in science, the curricula of students in the two faculties remained very similar. Accordingly, within the Mathematics Department, very little distinction has been made between Arts and Science students, and staff have not been allocated to any particular faculty.

The Professor of Mathematics in 1893 was William Jack (1834-1924). He was a Glasgow graduate, who had proceeded to Cambridge, where he attained the distinction in 1859 of being Fourth Wrangler in the Mathematical Tripos. He had had a varied career before being appointed in 1879 to the Glasgow Chair. He was successively an Inspector of Schools, Professor of Natural Philosophy at Owen's College, Manchester (now the University of Manchester), editor of the *Glasgow Herald* and publisher with Macmillans. He made no noteworthy contributions to mathematics as an academic subject, but he was an able administrator and acted as a new broom when he returned to the University.

His predecessor, Hugh Blackburn, also a Cambridge Wrangler, and a friend of Lord Kelvin, had resigned in 1879 because of increasing deafness and an inability to control his classes. Although a useful colleague to Kelvin who used him as a sounding-board for his theories, he seems to have become more interested in the early history of the University than in mathematics, and next to no records of the Mathematics Department have survived from his period of office, or from his predecessors.

Jack changed all that. He purchased seven leather-bound registers for use by the different mathematics classes. On the right of each student's name, there was a line of small boxes, each box corresponding to one week of the academic year. In each week and class there were normally four lectures and one tutorial (example class). Absences from lectures were recorded in each box by small strokes at 12, 3, 6 and 9 o'clock, and a central small circle indicated absence from a tutorial. The roll had to be called each day and the marking of absences was made easier because students sat in assigned places in alphabetical order, the women in the first few rows. This system was still in operation, though in a somewhat reduced form, in the 1960s and 1970s, but from 1972 onwards, students of either sex were allowed to sit where they wished. These early registers show that there were two other marks which could be entered to denote 'improper behaviour', namely an oblique stroke for 'condonable' and a cross for 'serious' misbehaviour, but these were very rarely used.

In the University *Calendar* for 1893-94, the names of George A. Gibson (1858-1930) and Daniel Lamont (1864-1950) are listed as Assistants in the Mathematics Department, indicating that the complement of the Department was only three. However, the more elementary teaching of the classes for women students, which were held outside the campus in Queen Margaret College (formerly North Park House and the Scottish headquarters of the BBC) was conducted by Peter Alexander, a tutor who was specially employed for this purpose and advertised his services in the advertisement section of the University *Calendar*.

In 1895 Gibson became Professor of Mathematics at the Glasgow and West of Scotland Technical College (now the University of Strathclyde) before returning to the University to succeed Jack on his retiral in 1909. Jack's other Assistant, Daniel Lamont, left in 1896 to enter the Church. He was ordained in 1900, served in various charges and became Professor of Practical Theology in New College, Edinburgh, in 1927. In 1936 he was appointed Moderator of the General Assembly of the Church of Scotland. In the early days the work-load of University staff was considerable, and must have left little time for research. Assistants had an additional burden, as it was customary for the professor to retire to the country during the summer term, leaving them to conduct classes for Honours students.

Jack, who was described by one of his staff , R. J. T. Bell, as "a charming old fellow", continued in office until his 75th year of age before seeking permission from the University Court to resign. It has been suggested that the reason why he did not retire earlier was that, while his predecessor Blackburn was alive, he might not qualify for a pension. However, there is no evidence for this and, in fact, Blackburn died some months after Jack submitted his application for a retiring allowance, as it was called. This was on the grounds that "the physical and mental energies necessary for the satisfactory conduct of the classes" for which he was responsible were "no longer at his disposal". A more likely reason for his seeking to retire at this time is that he had completed 30 years service as a professor in Glasgow, and so had reached the point at which his pension attained its maximum value, namely two-thirds of his salary. It is clear that, at that time, the provisions of the relevant Ordinance were very generous; they provided him with an annual allowance of more than a thousand pounds a year, a considerable sum in those days.

His successor Gibson did original work on problems in mathematical analysis. In his early days as an Assistant to Professor Jack, he had spent some time studying mathematics in Berlin, where he attended lectures by Kronecker and Weierstrass. Over the years he wrote a series of useful mathematical textbooks. Only two of these are mentioned here. His *Elementary Treatise on the Calculus* (Macmillan, 1901, 1906) was one of the better books on the calculus at a time when the standard works on the subject in common use in British colleges and universities left much to be desired in the way of mathematical rigour; it was only after the second edition of G. H. Hardy's *Pure Mathematics* (Cambridge, 1914) appeared that standards improved. Gibson's *Advanced Calculus* (Macmillan, 1931), a textbook aimed at more advanced students, appeared a year after his death.

Gibson had a great interest in the history of mathematics and published several papers on the subject. He retired in 1927 and, in the following year, his friends and students contributed to a fund for the endowment of lectures on this subject to be held at intervals in the University, and to be known as the George A. Gibson Lectures. This was, for its time, an imaginative action, and has helped in the resurgence of interest in the history of mathematics that occurred after the last War. The first Gibson Lecture was delivered in 1933 by Albert Einstein under the title *The origins of the general theory of relativity.*

The next holder of the Chair was Thomas Murray MacRobert (1884-1962), who was a graduate of Glasgow and Cambridge, and had been appointed to an Assistantship in the Glasgow Department in 1910, and to a Lectureship three years later. MacRobert was the author of numerous papers on the theory of special functions, on which he was an expert. He continued to be active in research until just before he died. Much of his work was connected with the properties of *E*-functions; these functions, which were invented by him, contain as special cases several of the well-known functions of mathematical physics and make it possible to give a unified treatment of the subject. His books on *Functions of a Complex Variable* (Macmillan, 1917) and *Spherical Harmonics* (Methuen, 1927) were widely used by students, and he also collaborated with other writers on books on Bessel Functions and trigonometry, He was responsible for the second revised edition of T. J. I'A. Bromwich's classical textbook *Infinite Series* (Macmillan, 1926). The value and popularity of these works are attested by the number of different editions in which they have appeared.

He was a friend and helper, not only to generations of Glasgow students, but also to many mathematicians from India and the East, who benefited by the willing assistance he gave them in their work without ever expecting acknowledgement. He was one of the original founders in 1927 of the Glasgow Mathematical Association. This society of schoolteachers and university staff was originally set up to support the continued teaching of Euclidean geometry in schools, but now had wider aims and is affiliated to the larger Mathematical Association of Great Britain.

When MacRobert was a member of the University Court he was successful in persuading that body to give financial support to a new mathematical periodical, which was known as the *Proceedings of the Glasgow Mathematical Association* during its first 16 years; the title was changed in 1967 to the *Glasgow Mathematical Journal*, since the original title had become misleading, the connection with the Association being tenuous. The Journal has flourished over the years and has been of considerable benefit to the University Library because of the number of exchanges arranged with other mathematical journals, and it has enhanced the reputation of the Mathematics Department.

MacRobert retired in 1954; he was succeeded in that year by the present writer, Robert Alexander Rankin (b. 1915), who retired in 1982. For financial reasons, no successor was appointed until 1989, when Robert Winston Keith Odoni (b. 1947) came from Exeter to fill the Chair . However, three new Chairs had been founded in 1955, 1966 and 1983. The first of these, named after Robert Simson, was held by Ian Naismith Sneddon, F.R.S. (b. 1919) from 1956 until he retired in 1985; it then fell into abeyance until 1992, when Brian Straughan (b. 1947) was appointed. The second is named after Sir Thomas Muir, F.R.S. (1844-1934), a former Assistant to Professor Blackburn, who in 1892 went out to Capetown at the invitation of Cecil Rhodes to be Superintendent-General of

Education in Cape Colony, and was the author of the monumental four volume *History of Determinants* (Macmillan, 1906-1923); this Chair is held by Walter Douglas Munn (b. 1929), who was appointed in 1973. To the third Chair, named after George Sinclair, Raymond William Ogden (b. 1943) was appointed in 1984, and he is currently Head of the Department.

In addition, it has been possible since 1966 to promote lecturers of distinction to Titular (ie, personal) Chairs. The first of these appointments in Mathematics was that of John Hunter (b. 1922), who served from 1976 to 1986. Others are still in post, namely Jeffrey Ronald Leslie Webb (b. 1945) appointed 1987, Kenneth Alexander Brown (b. 1951) appointed 1990, and Patrick Fleming Smith (b. 1943) appointed 1991.

The first Reader in the Department was Henry Richard Dowson (b. 1939), who was appointed in 1976 and retired in 1990. There are currently two other Readers in post, namely Stephen James Pride (b. 1949), appointed in 1987, and David Ross Fearn (b. 1954), appointed in 1992. All these new posts have added greatly to the strength and reputation of the Department, and have enabled the University to retain the services of able people, who otherwise might have departed to other universities.

This seems an appropriate place to mention former students and members of staff of the Department, who played a prominent role in the University or elsewhere. In the early years there were fewer universities in Britain, and it was therefore sometimes difficult for able lecturers to obtain advancement by moving to other universities. Thus Robert Franklin Muirhead (1861-1941) was a graduate of Glasgow and Cambridge universities and spent some time at the University of Göttingen. He held lectureships in Glasgow and Birmingham for brief periods and tutorships in various cities, but never held any permanent position worthy of his talents. He wrote papers on special functions; some of his work on convexity, which is of interest to statisticians, has recently come into greater prominence. Also worthy of mention is John Dougal (1867-1960), who did important work on hypergeometric and other special functions, and became scientific adviser to Blackie & Company.

However, others such as Horatio Scott Carslaw (1870-1954), Robert John Tainsh Bell (1877-1963) and James Morton Hyslop (1907-1984) emigrated to the colonies to take up Chairs in Mathematics. Carslaw was a handsome man and was known as 'Pretty Carslaw' to his female students, who showed their appreciation by laying a flower buttonhole on his desk before lectures. He went to Sydney University in Australia. Bell left Glasgow to go to Dunedin, New Zealand,

and Hyslop went to Witwatersrand, South Africa, and later became Vice-Chancellor of Rhodes University. They were all well-known for their books on various branches of mathematics.

Of other Lecturers who taught in Glasgow until their retirement, the most notable were Neil McArthur (1883-1951), John McWhan (1885-1943), William Arthur (1894-1979), Richard Alexander Robb (1901-1977), Robert Pollock Gillespie (1903-1977), and Thomas Smith Graham (1905-1974). They were all graduates of the University of Glasgow. McArthur was a member of staff from 1906 to 1948, and was the co-author of a widely used textbook on algebra. McWhan served on the staff from 1910 to 1943, and was concerned mainly with the teaching of engineers, for whom there were then three classes in mathematics, a fourth being introduced in the early 1950s. He was a Ph.D. of Göttingen University, where he had worked on thermodynamics.

The remaining four were Senior Lecturers when I came to Glasgow in 1954 and, in different ways, were towers of strength. They were responsible for different areas of work in the Department, and were excellent teachers. William Arthur, M.C., a man of military bearing, was the oldest and outlived the other three. He was responsible for the teaching of engineers, but also served for 17 years as Mitchell Lecturer in Statistics. This Lectureship was founded and endowed in 1925 by Sir George Arthur Mitchell (1860-1948), a distinguished mining engineer. After his retiral Arthur spent two and a half years teaching mathematics in Bethany College, West Virginia, and returned to Glasgow with an award from the student body as the best teacher on the campus.

Richard Robb began his academic career as an astronomer, and spent periods doing research in Glasgow, Michigan and Lund. Between 1935 and 1937 until Professor W. M. Smart was appointed to the Chair of Astronomy, he acted as Head of the Department of Astronomy. His work on astronomy led him into statistics. He succeeded Arthur as Mitchell Lecturer in 1944, and in 1950 became Head of the newly established sub-department of Statistics.

In the years that followed it was not only to astronomical questions that he applied his knowledge of statistics. A constant stream of research workers in every branch of science, and particularly in medicine, came to him for advice and help, which he gave willingly without asking for recognition or even acknowledgement. His friends sometimes suspected that quite a number of articles and dissertations that had won their authors' praise and preferment represented work essentially done by Dr Robb. The faculties of Arts and Science also benefited from his advice and from his analyses of the progress of students in relation to their entrance qualifications. All this work, together with the supervision of research students, was undertaken cheerfully by him without complaint, although it left little room for the prosecution of his own scientific interests.

Robb never thrust himself into the limelight, and some of his colleagues may not have realised that in his day he was a distinguished athlete. He participated in amateur athletics throughout his active working life. In 1928 he was a member of the British sprint team in the Olympic Games held in Amsterdam and he held several Scottish sprint titles.

Dr Robb retired in 1966, and this seems an appropriate place to mention that in that year a separate Department of Statistics and a new Chair of Statistics were established. The first holder of this Chair was Samuel David Silvey (1924-1990) who was a Glasgow graduate and had been a Lecturer in the Department of Mathematics. He moved in 1961 to the University of Manchester where he held a Readership in Statistics. His most lasting contributions to the subject are his two books *Statistical Inference* (Chapman & Hall, 1975) and *Optimal Design* (Chapman & Hall, 1980). Professor Silvey brought with him from Manchester John Aitchison (b. 1926) to a Titular Professorship. He left in 1977 to go to the University of Hong Kong as Professor. Professor Silvey retired in 1981 and was eventually succeeded in the Chair in 1988 by Professor Donald Michael Titterington (b. 1945). The Department of Statistics has increased appreciably in size since its foundation and now is one of the UK's largest centres for academic statistics, with 13 members on its academic staff. It runs a very popular single Honours degree as well as three combined Honours degrees.

In addition it houses, in the Boyd Orr Building, the Databases Unit, consisting of 19 members and led by Dr Ian Ford (b. 1951), currently Reader and Head of the Statistics Department, as well as a new group funded by the University Funding Council under the Teaching and Learning Programme directed by Sir William Bowman (b. 1955), a Senior Lecturer. A further imminent expansion is the creation of the Robertson Centre for Biostatistics, through which a new senior lecturer will direct statistical support for medical and biological research across the University.

The Department of Statistics enjoys an outstanding reputation for research, the University being one of the two institutions nationwide that obtained the top rating of 'five' for Statistics in both the 1989 and 1992 Research Assessment Exercises conducted by the Universities Funding Council.

Two men with Glasgow connections, who made substantial contributions to statistics, although neither was a member of the academic staff of the Mathematics or Statistics departments, may be mentioned here. The first was Sir Donald Alexander McAlister (1854-1934), who was Principal of the University from 1907 until 1929, and Chancellor thereafter until his death. He studied first at Cambridge and was Senior Wrangler in 1877, before turning to medicine as a career. He was the author of an important paper on what is now called the lognormal distribution.

The second was John Brownlee (1868-1927), who graduated in 1889 with first class Honours in Mathematics and Natural Philosophy, and then went on to study medicine. He eventually became Director of Statistical Services of the Medical Research Council and was the author of 80 papers on statistical subjects. It is interesting to note that he was probably the first person to lecture on statistics in the University; during the period 1910-12 he was invited to deliver a course of lectures on statistical methods within the Department of Mathematics.

I return to the remaining two senior members of the Department of Mathematics mentioned earlier, namely Dr Gillespie and Dr Graham. In the 1950s, their particular responsibility was the efficient running of the large Ordinary and Higher Ordinary classes, although both of them lectured to Honours students as well. Gillespie was an analyst and Graham was an algebraist and they both had Cambridge Ph.D. degrees. They were close friends, but of widely different characters, and liable to take entirely different views on such matters of examining and the award of class tickets. When he came in 1954, their new professor, who had the final responsibility in such matters, found it best to steer a middle course between them.

It would be true to say that, before the end of World War II, the mathematical curriculum had been somewhat biased towards those parts of mathematics based on analysis. Graham, who had spent two years in Princeton working with L. P. Eisenhart, fought successfully to effect a more even balance between analysis-based and algebra-based subjects, and it was due to him that Hans Zassenhaus, the distinguished German algebraist, spent a rehabilitating year in Glasgow immediately after the War. During his stay he provided much research stimulus.

The training of young assistants was largely in the hands of Gillespie and Graham. This training was carried out by them, in their earlier years at any rate, with a strictness and attention to detail that is no longer fashionable, although it was very much in the tradition of the Department. That it was extremely efficacious in producing excellent and meticulous lecturers cannot be doubted.

As has been indicated, in the early years of the Faculty, the mathematics classes available to students were identical with those in the Faculty of Arts. However, in October 1921, a new Faculty of Science Ordinance came into operation. The new regulations enabled Honours students to offer Mathematics alone as their single Honours subject, and did not provide for combined Honours in Mathematics and another subject (usually Natural Philosophy). However both Gibson and MacRobert discouraged students from offering single Honours, as they believed that an Honours degree in Mathematics should have an appreciable applied content, so that this new option of single Honours was discouraged, and combined Honours in Arts remained the standard curriculum for Honours students. However, at the beginning of World War II, the regulations for call-up of students were more favourable for Science than for Arts students, and consequently a combined Honours B.Sc. degree was offered and taken by many students. It became evident later that the granting of a B.Sc. with combined Honours violated the provisions of the Ordinance, and at a later stage, an enabling clause had to be approved in order to validate the degrees illegally awarded. It was not until the early 1950s that a small proportion of students regularly offered single Honours in Mathematics. Since then the numbers doing so have increased. It is now the option chosen by the majority of Honours students in the Mathematics Department. Similarly, most Honours students in the Statistics Department are single Honours students.

A curious feature of the new regulations was that it allowed an arts student who had graduated with Honours in Mathematics and Natural Philosophy, to obtain an Ordinary B.Sc. after one year's further study. This provision was frequently used in the 1920s, but was omitted when new regulations were introduced in 1967.

The spread of subjects available for study by single Honours students has increased enormously since the 1950s. As Ordinary and Higher Ordinary courses covering applied mathematics were taught by the Department of Natural Philosophy (now under the new title of Physics & Astronomy) it had been thought that more advanced work in that subject could not be taught in the Mathematics Department. However, agreement was reached which enabled the Department to provide Honours courses in all branches of applied mathematics. It should also be borne in mind that 50 years ago the term applied mathematics was used in the narrow sense of those parts of mathematics of a more practical kind needed by students of physics, whereas nowadays the distinction between pure and applied has diminished and applied mathematics can include subjects like combinatorics and coding theory.

Another change has been the continuing increase in the number of research students in Mathematics coming to Glasgow from all over the world and the massive amount of research published by staff and students. The pattern before the 1939-1945 War was that research students rarely came to Glasgow from other universities, and a handful of bright Glasgow students went elsewhere, mainly to Cambridge, to do research. New staff were largely recruited from these when they had completed their doctorates.

The choice of courses in Mathematics for Ordinary degree students has also been widened. Formerly far too many students embarked on the Honours course and found themselves out of their depth by their final year, although they were by no means without ability. Largely through the efforts of the Mathematics Department, a new course beyond the Higher Ordinary level, and called Advanced Ordinary, became available in Mathematics and other science subjects in 1963. This became a very useful course for those taking up teaching as a career, and helped to remedy the dearth of schoolteachers of mathematics.

A hundred years ago the Mathematics Department was housed in the main building, and most lectures were given there, in the Mathematics Class Room, which was a large bare lecture room situated above the rather more resplendent and panelled Humanity Class Room. The benches on which the students sat had been brought from the Old College in the High Street when the University moved to Gilmorehill in 1870. The mathematical benches were made from strong but inferior wood, which, by the time they were replaced, had become full of splinters. These had a deleterious effect on the women students' stockings. The benches were replaced eventually by tip-up seats, which are more comfortable as seats but more easily broken.

Next door to the Mathematics Class Room was the Mathematics Sideroom, a small room containing registers and other useful material, where the staff met between lectures. On its further side was the Astronomy Class Room, which was used at 10am for lectures to students of mathematics in the Intermediate Honours class.

Before the University Chapel was built, the Western Quadrangle opened out into the Professors' Square. But after that, new lecture rooms and staff offices were built on either side of the Chapel and became available to Mathematics and other departments. This increased enormously the accommodation available to the Department. In time, with growing student and staff numbers, this in turn became inadequate and extra rooms had to be found in University Gardens and Lilybank Gardens.

The case for a new building, which would house all the staff of the Department and provide extra lecture rooms, was strong. In October 1969 the staff moved into the new Mathematics building in University Gardens and gave up its accommodation elsewhere. The building is shared with the statisticians and astronomers, but over half the building is used by the Mathematics Department. Nothing ever remains static, however, and, as might be guessed, the building cannot now house all the Mathematics and Statistics staff and research students, and extra accommodation has had to be found elsewhere.

The increase in the size of the University over the years can be measured by the numbers of academic staff listed in the annual *Calendars.* Thus in Mathematics the number of academic staff has grown from three in 1893 (as already mentioned) to seven in 1920, 18 in 1956 and 37 in 1993. Before 1950, research students were very thin on the ground, but after that there was a gradual increase, largely due to the increasing number of professors and readers, and in the current year there are 24 research students. There has been a corresponding increase since 1950 in the number of visiting research workers from other countries and universities.

The number of female academic staff has, unfortunately, not grown at the same rate. The first woman to join the staff of the Mathematics Department was Miss Janet Fulton Rodger. She graduated with Honours in Mathematics and Natural Philosophy in 1919, was appointed to an Assistantship the following year and promoted to a Lectureship in 1923. She resigned in 1925 on her marriage to Thomas Sharp, a minister in Larkhall and later in Paisley. She died about 1970.

The numbers of female staff have never been as high as during the session 1953-54, when there were five, namely one lecturer and four assistants. At that time it was the custom to appoint to assistantships, not only those who hoped to make careers as university lecturers, but also bright graduates who intended to go in for schoolteaching; those in the latter category, of whom a high proportion tended to be women, were allowed to go directly into schoolteaching after they had completed their time as assistants, provided they had attended afternoon classes at Jordanhill College of Education during their first two years on the staff of the Mathematics Department. In more recent times, undoubtedly the main reason for the small number of female staff is the corresponding small proportion of women applying for vacant university posts, but one cannot help feeling that this excuse is perhaps just 'passing the buck'.

In conclusion, it is perhaps worth mentioning that the University possesses portraits of a number of its professors of Mathematics, namely Simson, James Thomson (1786-1849 - father of Lord Kelvin and the predecessor of Hugh Blackburn), Jack, MacRobert and the writer. There is also a marble bust of Blackburn.

Acknowledgements
I am grateful for help received from Mr Michael Moss, Dr Daniel Martin, Professor Ray Ogden and Professor D. M. Titterington.

Robert A. Rankin is Emeritus Professor of Mathematics at the University of Glasgow.

Natural Philosophy

Ian S. Hughes

The study of natural philosophy has been a cornerstone of the University of Glasgow since 1577. Indeed, among the early original fields of study in the University, it is the only one which has maintained a leading role into the present day. That this has been so is due primarily to the fundamental nature of the subject and to its rapid and continuing development. The hundred years since the foundation of the Faculty of Science in the University have seen the most dramatic developments in the history of the subject, and in this essay I attempt to contrast the state of knowledge and outlook in this area at the start and end of this period and to describe how the Glasgow Natural Philosophy Department, now the Department of Physics & Astronomy, has shared in this history.

Kelvin and the 1890s

When the Faculty of Science was instituted in 1893, physics in Glasgow was in a particularly healthy condition since the Chair of Natural Philosophy was held by William Thomson, at that time just recently (1892) enobled as Baron Kelvin of Largs, the first scientist ever to be appointed to the House of Lords. Kelvin had been appointed to the Chair in 1846 at the early age of 23, surely one of the wisest appointments ever made by the University of Glasgow. Initially Kelvin had lectured twice each day on five days each week for some six months in the year as well as being responsible for the experimental physics laboratory. By 1893 J. T. Bottomley who had acted as Kelvin's deputy since 1874 carried a substantial part of the teaching load but Kelvin continued to lecture twice a week. In the effort to envisage the University one hundred years ago it is of course critical to remember that 'departments' as we know them did not exist. The Professor dominated and controlled teaching and research. In Natural Philosophy at that time, apart from Kelvin himself, there were James Thomson Bottomley (Lady Kelvin's nephew), himself a D.Sc. and a Fellow of the Royal Society, who held the appointment of Arnott and Thomson Demonstrator in Experimental Physics, an Assistant, Magnus McLean, (a Fellow of the Royal Society of Edinburgh) and in addition James Blyth who held an appointment as a Lecturer in Natural Philosophy and whose duty was to teach the young ladies at Queen Margaret

College, which had been absorbed by the University in 1892 but still provided separate classes for women.

The dominant position of the holder of the Chair meant that to have a professor of such distinction as Kelvin was an outstanding asset both to the students who benefited from his teaching and to the reputation of the University. In 1893 Kelvin was aged 70 and had held the Natural Philosophy Chair for 47 years. He was undoubtedly the most distinguished physicist in the UK with a world wide reputation as one of the most outstanding physicists then alive. Although most physicists would probably agree in retrospect that Kelvin's most important contributions to the subject were in thermodynamics and had been made some 40 years earlier, he was still extremely active and continued to publish papers at an astonishing rate right up to his death 14 years later in 1907. In 1893 he published five papers (some in several versions in various journals), mainly on piezo- and pyro-electricity and elasticity in crystals, as well as giving prestigious lectures and addresses (including the Presidential Address to the Royal Society of which he had been a fellow since 1851 and president since 1890), and there were some 97 further papers to come.

A study of Kelvin's later work gives us not only a rather clear idea of the kind of physics taught at Glasgow in the 1890s, but also of the classical conception of the physical world held by the 'establishment' in physics at that time before the revolution in the subject generated by the discovery of the electron, radioactivity, the nuclear atom, relativity and, most of all, quantum mechanics. This conception is exemplified in the 'Baltimore Lectures', a set of 20 lectures which Kelvin was invited to give at Johns Hopkins University in Baltimore in the fall of 1884. The lecture series was almost in the nature of a 'master class' in that the audience included a number of distinguished physicists well known to us today, such as Lord Rayleigh, Michelson and Morley, as well as other very able professionals from the USA, Canada, Japan and Europe. Kelvin modified and adapted the lectures as the course proceeded and he developed ideas stimulated by the response of the audience. One might expect that the severely classical position taken by Kelvin which I shall describe below might have been modified in the following two decades, but in fact the lectures were published by Cambridge University Press in 1904, after revision by Kelvin, a revision made with the knowledge of subsequent developments but which did not fundamentally change his mechanistic position.

Kelvin strove to the end to understand nature in terms of mechanical models. One of his most quoted statements was made in the Baltimore Lectures:

"the test of 'do we or do we not understand a particular point in physics?' is 'can we make a mechanical model of it?'" and "I never satisfy myself until I can make a mechanical model of a thing. If I can make a mechanical model I can understand it. As long as I cannot make a mechanical model all the way through I cannot understand; and that is why I cannot get the electromagnetic theory". These statements of course raise fundamental questions still unanswered at the present day. What do we mean by 'understanding'? Most physicists would probably not be finally satisfied even by a complete set of equations predicting the development of the physical world, but the inadequacy of mechanical models was already evident for the really open minded, in Maxwell's theory of electromagnetism first published in full in 1873. Physicists now generally feel that they do 'understand' electromagnetism in that the concepts are completely familiar to them, and in that calculations in quantum electrodynamics can be made to almost any desired precision and the result of such calculations verified by experiment to one part in 10^8. Whether such comfortable familiarity and predictive success constitutes true understanding is open to debate.

Kelvin chose as the subject of his Baltimore Lectures 'Molecular Dynamics and the Wave Theory of Light', that is, the interaction of light with matter. We should recall that this was the age of the 'luminiferous ether' conceived by many including Kelvin as a kind of elastic solid pervading all of space. Optics was already a very highly developed science from the experimental point of view so that any model had to account for not only reflection, refraction and dispersion but also for phenomena such as optically active crystals, double refraction and the Faraday rotation as well as fluorescence, phosphorescence and line spectra. The rotational and magnetic effects presented particular problems for the classical ether.

Kelvin devised sophisticated mechanical models including molecules consisting of massless concentric spheres linked internally by springs connected to a central mass and embedded in the ether. This era has even been described as the 'high baroque' of the mechanical model. Kelvin's mathematical facility (he had been Second Wrangler at Cambridge and moved easily in mathematical application) allowed him to analyse these complicated models. Despite heroic efforts however, Kelvin recognised that he had not developed a satisfactory model to account for the wealth of phenomena in the interaction of light with matter.

In the light of later discoveries of course it is clear that Kelvin lacked some of the essential ingredients for a satisfactory model - in particular knowledge of atomic structure and quantum theory. More significant is the fact that, unlike some of his younger contemporaries, Kelvin was not easy with Maxwell's electromagnetic theory and failed to appreciate its strength and sufficiency. In the end of course, the ether faded away as its conception as an elastic solid required increasingly bizarre properties to remain consistent with experiment, and as realisation spread that it was an unnecessary artefact.

The summary undergraduate syllabus and the examination papers of the 1890s reflect the classical physics of the time. What was the Ordinary Class (the term only came into use in 1900-01) met at 9am as it does today. The subjects of study were dynamics, properties of matter, sound, light, heat, magnetism and electricity. The Higher class met at noon for 'Mathematical Demonstrations and exercises in Kinetics and Statics and a course on some special subject of Mathematical Physics'. The 'special subject' varied from year to year and Honours students were advised to take the Higher class in two consecutive years.

Prior to the institution of the Faculty of Science natural philosophy was an Arts subject on the same footing as logic and moral philosophy. In the words of Lord Bacon it was an instrument for "that improvement of the understanding which results from the cultivation of natural knowledge, and that elevation of mind which flows from the contemplation of the order of the universe". Many of us today believe that the study of physics is still an instrument for the accomplishment of these lofty ideals. One of Kelvin's great achievements however was surely that he demonstrated through many examples, of which the transatlantic cable was at the time possibly the most spectacular, the power of physics to generate technologies which have a profound effect on style and quality of life. This aspect of Kelvin's work was a feature of his teaching as much as it was of his own original research. Despite the practical aspect of the course however, it seems that only a few of the physics undergraduates went on to apply their knowledge in a material way partly due to the lack of such a tradition and partly through lack of opportunity. Kelvin's world wide reputation attracted a steady trickle of foreign students: Japanese, Italian, Spanish and German, to the Department. In addition the contribution made by Natural Philosophy to the education of the engineers in the University should not be underestimated although this became formalised in separate classes only two years after Kelvin's retiral from the Chair of Natural Philosophy.

Kelvin retired from the Chair of Natural Philosophy in 1899. As already remarked his retiral did not mark the end of his active involvement in physics. He matriculated as a student, this time a research student, in the University of Glasgow for a second time, 65 years after his first matriculation and aged 75.

Kelvin's immense prestige and strong opinions must have profoundly influenced the development or lack of it in Natural Philosophy in Glasgow not only until his death in 1907 but much beyond. Even among outstanding people it is rare to find the ability to accept radical new ideas at an advanced age and Kelvin with his fundamentally mechanical conception of nature was no exception to this pattern. As Stein (1) has pointed out: "Kelvin's view of physics was . . . curiously stiff and hidebound . . . the advances he celebrated in the preface of 1904 (to his Baltimore Lectures) . . . have left no trace on the physics of our century whereas . . . work from before the time of the lectures up to the 1904 publication, which had achieved the greatest advance in the subject of the lectures and laid the groundwork for the future was passed over in the lectures and in the published volume".

To succeed a man of such distinction as Kelvin must have been difficult, particularly since Kelvin had not created a strong 'department' with momentum to carry on his work. Kelvin had largely worked on his own, although in the 1890s he published a number of papers with his Assistants Magnus McLean and Alexander Galt. J. T. Bottomley had joined Kelvin as a Private Assistant in 1870 and worked on a variety of studies in heat and electrical engineering. Joint publications with Kelvin however are very few and Bottomley's major contribution appears to have been in lecture room, teaching laboratory and administration. The fifth member of staff, J. Blyth, appears to have been concerned only with teaching in Queen Margaret College for which he was paid £100 p.a. (Kelvin's university salary, exclusive of fees, was £1,252 p.a., but the most highly paid professor was G. Clelland of Anatomy at £1,645.) Bottomley, McLean and Galt all apparently retired or moved when Kelvin himself retired in 1899, only Blyth continuing into the new regime until 1907.

The early years of the 20th century

Kelvin's successor in the Chair of Natural Philosophy was Andrew Gray, F.R.S., a classical mathematical physicist. Gray had been a student of Kelvin's (M.A. 1876) and had acted as Kelvin's personal secretary and Assistant for a decade before being appointed to the Chair of Physics at the University College of North Wales at Bangor. The number of undergraduates in the Natural Philosophy class had fallen sharply through the 1890s from 234 in 1892-93 to 82 in 1897-98 and 92 in 1898-99 (figures 1 and 2) though the Higher/Honours classes had risen in the first half of the decade to a fairly constant number of 20-30. Kelvin is often credited with the establishment of practical teaching laboratories in

Glasgow and in leading the field in this respect in the UK (although he himself never made such claims). In fact a practical laboratory existed in Glasgow long before Kelvin was appointed to the Chair. The laboratory was however badly run down when Kelvin arrived and, recognising its importance and unsatisfactory condition, he persuaded the University to grant quite substantial funds for its refurbishment. There was in his day no clear separation between the teaching laboratory and his own research activity, and undergraduates found themselves assisting in Kelvin's original experiments. This healthy and instructive state of affairs has relatively recently been reintroduced to the Honours teaching laboratories in a small way in the form of undergraduate 'projects'. In the 1890s however, the laboratory was not a mandatory element of the Natural Philosophy class. In the 1893-94 session for instance, 220 men and 15 women studied 'first year' natural philosophy and eight men attended the Higher class, but only 15 students were enrolled for the laboratory class. In addition to these numbers we must not forget that an important duty of the Natural Philosophy 'Department' was to teach elementary physics to medical students - a class varying from about 100 to 130 students. This function of the Department continued (and expanded to include dental and veterinary students) until the reorganisation of the medical curricula in 1973.

The new Professor, Andrew Gray, undertook a major revision of the teaching arrangements in the Department, apparently intended to restore the student numbers, mainly in the first year class, to improve the Honours curriculum and to make the laboratory an integral and essential part of the course. New staff were recruited though the total number in the period before the 1914-18 War was always small, varying from four to seven including Gray himself. In the 1900-01 session a special Higher class in 'Dynamics etc.' (shortly to be known as Higher mathematical class section B) was instituted primarily for the engineers and a tutorial or 'problems' class was started, meeting twice a week. The first year Natural Philosophy class is for the first time referred to as the 'Ordinary Class'. The Higher mathematical class continued as did the class in 'Electricity Pure and Applied' which had been taught by Magnus McLean in the 1890s. The Professor once more took the main part in teaching the pure natural philosophy classes with other 'lecturers' teaching the engineers and medical classes. At this time 'assistants' were primarily research posts while 'lecturers' were appointed for a period not exceeding five years (reappointment was allowed), usually with a specific remit as to courses to be taught. Lectureships and assistantships or demonstrator posts could be, and often were, held by the same person at the same time. Only the professors apparently had long term tenure!

The summary syllabus for the Ordinary natural philosophy course did not change between 1895-96 and 1905-06: dynamics, properties of matter, sound, light, heat, magnetism and electricity. Not surprisingly for a subject dealing with eternal verities, the summary in fact changed only little in the next 50 years with the explicit inclusion of wave motion by 1946-47 and atomic theory by 1956-57. The summary, of course is only the barest description of the course content and there is no doubt that there were substantial changes in emphasis and the way in which the topics were treated, at least over the longer term.

The Higher Mathematical class alternated its syllabus in a two-year cycle and intending Honours students took the class twice in successive years. The syllabus remained severely classical, in a form which Kelvin would have found totally familiar, until the early thirties and, in some respects, even beyond that time. Kelvin's classic text book written with Tait of Edinburgh, *Thomson and Tait*, first published in 1872, was still a recommended text in 1907-08.

In Kelvin's day Natural Philosophy was accommodated in part of the south front of the Arts Quadrangle. Gray felt that this was overcrowded and ill equipped and one of his early tasks was to persuade the University to provide a new building. Kelvin was far from supportive of this project on the ground that the accommodation which had been good enough for him for 30 years was surely good enough for his successor. Nevertheless the project went ahead and a handsome stone building was built, superintended in detail by Gray himself who had been a stonemason in his youth. A feature which will be well remembered by the many thousands of students who have sat on its rather uncomfortable seats was the steeply raked large lecture theatre, with its balconies for demonstration rigs above the lecture bench, its spacious area for demonstration experiments and its high ceilings. The new building provided extensive teaching laboratories, research rooms, offices, workshop accommodation and storage facilities and was heated by a novel warm air circulation system. The cost of the project was £40,867-19s, funded by £20,000 from the Carnegie Trust, £20,000 from the University 'Equipment Fund' and the remaining £867-19s from the 'Building Extension Debt'. The accounts contain many interesting details: the elegant cast iron spiral staircase which many will recall was expensive for its day at £31-12-6, the cost of the trial pit was £48-4-1 but apparently failed to reveal that at the southwest corner of the building the old coal workings were only ten feet below the basement level. The inauguration, at a cost of £382-18-10, must have been quite a party.

There is a common belief that research in the Glasgow Natural Philosophy Department ceased after Kelvin's retiral and did not get under way again until the arrival of P. I. Dee in 1945. This is not in fact the case, but the reasons for such an impression having been created are easy to understand. Andrew Gray certainly saw research and a postgraduate research school on the German model as an important part of a good Physics department. The names of the research students begin to appear in the University *Calendar* for the session 1903-04; there were five of them including R. A. Houston, who later became a leading authority on optics and a member of staff, retiring only in 1948. The first lady research student, Miss Margaret Moir, registered in 1908-09. These postgraduate students were financially supported by (recently available) studentships: Houldsworth, Carnegie, Thomson scholarships. The research interests of the laboratory were mainly in the areas of magnetism and magnetic alloys, in gyroscopes and in optics. James G. Gray (the son of Andrew Gray) became an Assistant in 1903-04 and developed a lifelong passion for gyroscopes. He published a number of papers on this topic, worked on gyroscopic compasses for the Navy during the First World War and left the Natural Philosophy Department a collection of gyros which some of us have used in demonstrations up to the present day. By far the most prolific in publication was R. A. Houston who became an Assistant in 1907-08 (after taking a Ph.D. at Göttingen and spending a year at the Cavendish Laboratory in Cambridge working under J. J. Thomson) and a Lecturer in 1908-09 and whose interests spanned the whole field of optics, although he also worked on topics in magnetism. Houston published what became probably the most widely used undergraduate textbook on optics over several decades, *A Treatise on Optics.* The first edition appeared in 1915 and the seventh in 1938 with numerous 'new impressions', the last of which appeared in 1947. When asked how much money he had made from his book, Houston is said to have replied: "More than a Nobel prize".

In assessing the research contribution of the Natural Philosophy Department in the first half of the 20th century we must bear in mind that during all that period the staff (not including Applied Physics - see following page) never exceeded five, a number small even to teach a rapidly expanding subject. More important however is the fact that the research activity was in areas which were extensions of 19th century physics and did not touch on the revolutionary developments which were taking place elsewhere.

These revolutionary developments included J. J. Thomson's discovery of the electron in 1899, Rutherford's pioneering work on radioactivity around the turn of the century, Einstein's special theory of relativity in 1905 and the general theory in 1915, the Rutherford nuclear atom in 1911 and the whole subject of nuclear physics which sprang from that work, Bohr's atomic theory in 1913 and the development of quantum theory by Heisenberg and Schrödinger in the period 1925-28. Quantum theory was the key to future developments in all areas of microscopic physics: atomic physics, condensed matter (solid state), nuclear and particle physics. It required fundamental changes in the way physicists regarded matter and posed problems of acceptance and understanding never overcome by many older classical physicists. Kelvin's mechanistic model philosophy cannot encompass the concepts of quantum mechanics.

The fact that the Glasgow Department had no part in these new and revolutionary developments had two unfortunate consequences. On the one hand, its research activity appears as peripheral to what was the exciting mainstream during that period and, on the other, the staff were not able to include satisfactory courses on these important new developments in their undergraduate courses.

Applied physics

In 1921 the University took a decision to create an independent department of Applied Physics with responsibility for teaching all the service courses previously provided by the Department of Natural Philosophy. James Gray, son of the Professor of Natural Philosophy, who had been a Lecturer in the Department since 1906, was appointed to the new (Cargill) Chair. When shortly thereafter yet another Gray (unrelated) rejoined the Department, this led to the irreverent description of the trio as "Father, Son and Holy Ghost". James Gray had taken a degree in engineering at Glasgow in 1903 before becoming an assistant in Natural Philosophy the following session. Two lecturers transferred to the new department leaving only the Professor, Andrew Gray, and two lecturers in Natural Philosophy. Shortly thereafter three further assistants and a teaching fellow were appointed to make five staff in each department, a level which remained practically unchanged until after the Second World War.

The Department of Applied Physics had a substantial teaching load providing service courses for students of engineering, naval architecture, architecture, applied chemistry, metallurgy, pharmacy, agriculture, public health and medicine but no real applied physics was taught. The creation of a separate department of this kind appears to have been a retrograde step. Neither department had the resources of manpower and equipment to make a big research impact along with their teaching responsibilities. During H. A. Wilson's short tenure of the Natural Philosophy Chair in 1924-25 he carried out a major reorganisation of the departments (resigning only ten days after the change was decided) against the wishes of most of the staff, following which the two departments operated quite independently, within the same building, but otherwise with very little interaction. There were difficulties concerning division of the existing teaching apparatus and inadequate funds to provide enough new equipment. More fundamental than these problems however is the impossibility in my opinion of creating a vigorous department which provides only service courses, has neither a clear fundamental or vocational mission and lacks a research base.

James Gray retired from the Applied Physics Chair in 1935 and was succeeded by Thomas Alty who came to Glasgow from a Chair at the University of Saskatchewan in Canada. His main research interest was in all aspects of the physics of surfaces, a field in which he had first published while a graduate student at the Cavendish Laboratory and in which he continued to do elegant experimental work and to make significant contributions to theoretical understanding during his period in Glasgow. He was, not surprisingly, unhappy with the divided nature of the Department, and with the fact that his department had "no advanced students, . . . no research students, . . . no research apparatus" and described his efforts to deal with these problems in the years just before and after the start of the 1939-1945 war as a "devastating experience" (2). During the War Alty carried a heavy load of additional work with courses for Royal Artillery, Royal Engineers and RAF cadets as well research for the mine design department of the Admiralty.

The Applied Physics Department was discontinued in 1949 when Alty left Glasgow on his appointment as Master of Rhodes University College at Grahamstown, South Africa where he had a distinguished career as an academic administrator before returning to the UK as Principal of Birmingham University in 1963. During its 29 years of existence the Department undoubtedly served an important didactic function but in retrospect I believe its creation probably inhibited rather than promoted the development of physics in the University.

Taylor Jones and the period between the Wars

Andrew Gray retired from the Chair of Natural Philosophy in 1924. His successor was H. A. Wilson M.A., D.Sc., F.R.S. who came from the Rice Research Institute in Houston, Texas, to which he returned after less than one year in Glasgow, reputedly as a result of disagreements concerning personal and departmental financial provision. Edward Taylor Jones D.Sc., who was currently the external examiner, was appointed to the Chair in 1926 and served until the appointment of P. I. Dee in 1943. Taylor Jones is remembered by many Glasgow natural philosophy students for his lectures to the Ordinary Class which were famous for the demonstrations which he mounted especially when the subject was sound. He himself played the cello and used it to illustrate the lectures in the tradition of Kelvin with the French horn. This was an era of good natured rowdyism and, particularly at the first lecture when 'old lags' returned and on Daft Friday, there were some riotous scenes in the big lecture theatre. This rowdy tradition continued, particularly amongst Engineering students, until after the War when the standard of witty banter apparently declined and order was restored by less tolerant staff. A small research and postgraduate school continued during this period with a total of 16 research students between 1926 and 1943 all except one supervised by Taylor Jones himself, to the chagrin of R. A. Houston whose optics laboratory was probably the best equipped in the Department. One of Jones' special interests was electron diffraction and he produced some fine photographs with quite primitive coil-driven high voltage equipment. W. MacFarlane who had joined the Department as an assistant in 1926 and who later became Lecturer and Senior Lecturer, became responsible for the 300Mev synchrotron and served the Department and the University in many other roles, also worked on electrostatic lenses for electron microscopy, a technology which became important in later years.

P. I. Dee and the post-war renaissance

Already in 1943 the Principal, Hector Hetherington, had determined that after the rigours and restrictions of the War were over, the University should rebuild Natural Philosophy at Glasgow to become one of the nation's major Physics departments. To this end on the retiral of Taylor Jones in 1943 Philip I. Dee was appointed to the Natural Philosophy Chair, although he did not take up the appointment until his release from his work on radar at the Telecommunications Research Establishment (TRE) in 1945. P. I. Dee came to Glasgow with an exceptional reputation as both an experimental physicist and a leader of research projects. Like many of the post-war heads of UK Physics departments Dee had been a student and then a colleague of Rutherford in the Cavendish Laboratory in Cambridge. During the war he had played a key role as leader of a highly successful group developing radar at TRE at Malvern. Amongst a number of radar projects perhaps the most important was the development of an airborne radar system code named H2S which was first used in action in 1943. In addition to his reputation gained at the Cavendish and at TRE Dee had the advantage that his work during the War had brought him into contact with most of the leading UK figures in physics at that time.

The vital importance of physics in generating new and powerful technologies had been demonstrated in dramatic fashion during the War. Many examples could be quoted but the most spectacular are probably radar and the atomic bomb, the latter an achievement since regarded with mixed feeling by physicists as by others. The recognition of the practical importance of physics resulted in strong government financial support. The interest generated not only by the applications but by the fundamental pre-war discoveries and developments, which had by then become widely known, resulted in a generation of enthusiastic students. This was the start of what was possibly a uniquely favourable period for the development of Physics departments. The support provided for Dee to build a greatly enlarged and vigorous Department indicates clearly that the Principal was well aware that this was an opportunity to be seized. Dee certainly needed such support since overall facilities in Glasgow had been allowed to decline during the first half of the century. Symptomatic of the situation is the hardly credible fact that the Department did not enjoy an alternating current supply, the University being one of the last parts of the city to be on direct current.

There were four major elements in Dee's strategy to transform the Department from an undistinguished position to one of the foremost teaching and research institutions in the UK. These were a dramatic increase in teaching and research staff, a thorough review and reform of the undergraduate curriculum, installation of major research tools, in particular the 330 MeV electron synchrotron and the Cockroft-Walton High Tension accelerator (HT set), and a new building for research and teaching. The key element was without doubt the recruitment of a large number of young and enthusiastic staff.

There were six members of staff in post when Dee arrived, most of whom were nearing retirement and five of whom retired or moved within the following few years. By 1950, only five years after Dee arrived and prior to the major overall increase in the staff at Scottish universities which took place in the 1960s following the Robbins report, the total academic staff had grown to 33 (figures 1 and 2). Eight of these were research fellows, devoted full time to research and supported from external sources (Nuffield, ICI). Most of the new appointments were of graduates from universities other than Glasgow, many of them former colleagues of Dee at TRE. Exceptions were Glasgow graduates I. N. Sneddon, later to become Simson Professor of Mathematics in Glasgow, and S. C. Curran, later to become Principal of Strathclyde University. New appointments also included two from continental Europe: K. Guggenheimer and B. Touschek, who was later appointed to a Chair in Rome and became internationally distinguished for pioneering work on collider accelerators, now the key to high energy particle studies.

The experimental research programme was planned to concentrate on nuclear and what is now known as particle physics. The primary facility was to be the 330 MeV electron synchrotron. This choice complemented proton accelerators planned for Liverpool and Birmingham universities. The electron machine generated a beam of photons having an energy spectrum with a maximum of 330 MeV, well above the threshold for production of mesons predicted by Yukawa as far back as 1935 but discovered in cosmic rays only in 1947. The synchrotron was manufactured by Metropolitan Vickers Ltd and commissioned in the Glasgow laboratory by a small group led by Dr W. MacFarlane. First experiments were planned for 1951 but in fact the machine first operated for experiments only in 1954. These were early days in the development of such machines and the effort required to build and commission the synchrotron was seriously underestimated. Similar delays in commissioning marked practically all the post-war UK accelerator projects. The consequence of delay was that the most exciting and easier experiments had already been done by several US accelerators (in particular the synchrotron built by MacMillan at Berkeley which operated in 1948) before the Glasgow machine got under way. Nevertheless the Glasgow physicists (in particular J. R. Atkinson, E. Bellamy, W. Hogg, G. M. Lewis, J. M. Reid, J. G. Rutherglen, D. Stewart and W. S. C. Williams) carried out an important programme of work primarily on meson photoproduction near threshold and on nuclear photodisintegration, during the ten years after 1954.

Experimental physicists were not of course idle while waiting for the synchrotron. A small (30 MeV) electron synchrotron obtained from Harwell was used for nuclear photodisintegration studies with cloud chambers. A programme of work using radioactive sources generated valuable results in nuclear physics (for example what was at the time the best upper limit on the neutrino mass was measured by S. C. Curran and collaborators) as well as contributing to the development of sensitive detectors.

A major activity developed in the years shortly after Dee's arrival was the 1 MeV Cockroft-Walton accelerator (HT set). Dee had himself been responsible for the construction and commissioning of a 2 MeV Cockroft-Walton machine at Cambridge completed just before the War. He told the writer that many people outside Glasgow had questioned his decision to build an HT set in Glasgow, but it was a decision vindicated over a decade by a highly fruitful programme of studies of low energy nuclear reactions which provided some of the detailed nuclear spectroscopic data on which later models of the nucleus were built. The basic machine was bought from Philips of Eindhoven but a great deal of work to provide the ion source and commission a working accelerator was done by John Rutherglen (initially a Research Fellow and then Lecturer, Senior Lecturer and Titular Professor in Glasgow), E. Rae, R. D. Smith and colleagues, at very modest cost using a great deal of ex-War Department equipment, a source of cheap hardware which was vital to many projects in Physics (and also other departments) for many years after the War.

Dee believed that the development and perfection of instrumentation, in particular of particle detectors, were an almost infallible route to new discoveries. He was willing to support such work even if no application was apparent, and I believe, was often motivated by development of beautiful instrumentation more than by fundamental physics objectives. This is not of course to suggest that he was not mindful of such objectives or indeed that he was not conscious of the key role of theoretical physicists in both research and teaching. The folklore is that Rutherford was antipathetic to theoretical physics and that he imbued his students with this outlook. A closer study (3) of Rutherford's work and opinions reveals that this impression is much exaggerated, and Dee for his part took the opportunity of a vacancy in the Cargill Chair in 1948, when Professor Alty resigned, to persuade the Court to assign the Cargill Chair to Theoretical Physics. The first, and so far the only, holder of the theory Chair was John C. Gunn appointed in 1949. Dee was determined that the experimental and theoretical work should be parts of a unified department fulfilling complementary roles in research and teaching. This arrangement may seem only natural to readers in other disciplines, but in fact in a number of universities the departments are separate, although in some of these it has become clear that a marriage between the two elements is highly desirable. I believe the wisdom of having a united department has been borne out by the experience of the last 45 years.

An important feature of the new regime was the 'New Building'. The 'Old Building' of 1907 had served well when the Department had a small staff and a limited research activity concerned with small scale projects. The sharp increase in staff and research activity, the scale and complexity of the new work, the need for improved teaching laboratories, particularly for the Honours classes, in themselves required a major addition to the building. In addition the electron synchrotron required custom-built accommodation including stable, well shielded chambers for the machine and experiments, a control room, electrical supply equipment and crane facilities. All these requirements were addressed in a new building adjacent to and linked with the old Natural Philosophy building. Two problems nearly stopped the new building project before it even started. One problem was the extensive mine galleries which underlie the northwestern end of Gilmorehill but it proved possible to locate and fill the galleries under the new site. The other problem was the strong objection raised by professorial residents in The Square who at one stage threatened to go to law to stop the building. Fortunately the Principal managed somehow to deal with these objections without recourse to the courts although he did have to explain some delay on this account to the University Grants Committee (UGC).

The new building was designed by the well-known architect Basil Spence and was constructed in two stages, the first being the new research building including the synchrotron accommodation and a new lecture room, completed in 1951, while the second stage provided new teaching laboratories and a further lecture theatre, completed about 1959. The new building had the advantage that it was built before the restrictive austere norms of the University Grants Committee were laid down. The corridors are thus sufficiently wide and lined with terazzo panels, rather than narrow brick lined alleys. Its disadvantage lies in the idea that staff members should share their rooms not only with their postgraduate students but also with their experimental equipment, which in certain cases was noisy or even hazardous. The quite large 20' square rooms provided as a standard unit have proved rather inflexible. Most physics laboratories built since that time have been designed with large flexible experimental areas coupled with smaller offices for staff.

Important features of the new building were the new library and the 'Museum' on the top floor, both with a view south over the city. The Museum housed a display of part of the magnificent collection of historical equipment, mostly from Kelvin's laboratory. This collection benefited enormously from the attention of J. T. Lloyd, chief technician and later research technologist in the Department. Lloyd joined the Department in 1946, and in addition to his many other duties developed a deep interest in Kelvin and his instruments (many of which he refurbished), for which he made the definitive catalogue of the Glasgow collection. The Museum had a second function in that it served as the meeting place for tea at 4pm, an institution which Dee felt to be an important element of communication within the Department and which thrives to the present day. One of Dee's injunctions to new postgraduate students was that they were expected to come to tea and to attend all the colloquia.

The teaching arrangements and in particular the Honours curriculum were thoroughly revised. Dee broke the tradition that the Professor of Natural Philosophy should give substantial courses to the Ordinary class which was looked after for many years by J. McNeill. The only change which shows in the *Calendar* summary for the Ordinary class however is the addition of 'Atomic Theory' to the Kelvin list. Unlike more ephemeral subjects the elementary truths of physics are forever! With the increase in staff numbers it became possible to get rid of the practice of teaching the third and fourth year Honours classes together, with the curriculum alternating year by year, and to establish a well structured two year course for Honours students. Quantum theory became an important feature of the course. An Inter Honours second year course was set up as an alternative to the Higher Ordinary course and as the normal preparation for the third and fourth year course. A small group tutorial system was set up in which Honours students were assigned in ones or twos to staff members for weekly tutorial sessions. Whereas up to this time the normal practice for Honours students in Natural Philosophy was to take a joint Honours degree, usually along with Mathematics, the single Honours course was developed at this time and progressively overtook the joint Honours degree in terms of student numbers.

A feature of the Department in these days was the social atmosphere fostered by Dee. As well as the importance of afternoon tea already mentioned Dee held parties for staff and postgraduate students notable for a variety of slightly intellectual parlour games. The Physical Society party always included a feature mounted by the staff and included on different occasions memorable 'operatic', 'ballet' and 'orchestral' performances. The Physical Society outing normally featured a hill walk for the more energetically inclined. This event took place after the Honours examinations and before the results were announced, and was well attended by staff, postgraduates and undergraduates, some of the latter hoping for clues as to their class of degree. Staff student sporting contests were an annual event including cricket, golf - to which Dee was addicted - and occasionally football.

Research developments

The main thrust of the Department research was in nuclear and particle physics. Experimental research groups were distinguished primarily by the particle detection technique used, plus a Group led by J. Rutherglen using the HT set. Thus there was a Cloud Chamber Group led by J. R. A. Atkinson, a Counter Group in which academic staff members were initially S. C. Curran, A. Cockroft and G. Lewis and a Group using nuclear emulsions led by H. Muirhead who had come to Glasgow from Bristol where, as a postgraduate student, he was one of the discoverers of the pi meson.

The establishment of the Department on the world scene and the first operation of the synchrotron were marked in 1954 by the 'Rochester Conference' for that year choosing Glasgow as its venue. This conference is the major international meeting on particle physics, now held every second year. At that time, when particle physics had not emerged as an independent discipline, the conference also covered nuclear structure physics. The conference attracted many of the most famous names in the subject including Bohr and Heisenberg, Bethe, Weisskopf, Wheeler, Livingston, Leprince-Ringuet. It was also the scene for the first announcement of the Nobel prize winning theory of the 'strange' particles and the idea of the strange quantum number in a virtuoso dual performance by Gell-Mann and Pais which drew some incredulous laughter from some of the audience in the big lecture theatre. The Department will once again host this international conference in 1994, 40 years after the first occasion.

While the experiments on the electron synchrotron began to dominate the Glasgow programme a new and very powerful particle detector had been invented by D. Glaser in the USA in 1952 and was being developed into a sophisticated instrument ideally suited for the discovery and study of the new families of particles which were revealed in the following two decades. The bubble chamber in fact superseded both nuclear emulsions and cloud chambers for all but a few specialised applications and played a big role in particle physics for 20 years from the mid-1950s.

The Cloud Chamber Group in Glasgow moved into this technique and after some initial experiments built a 10" diameter propane chamber. In 1958 J. R.

Atkinson left Glasgow to take up a post at the Dounreay reactor establishment and I. S. Hughes returned from the USA where he had been working on a liquid helium bubble chamber studying K^- mesons at Berkeley, to join the staff and take charge of the bubble chamber work. The propane chamber was developed for use in experiments on the synchrotron, but the focus of the bubble chamber work was soon moved to the European international laboratory CERN at Geneva. When the CERN laboratory and UK membership thereof had first been proposed Dee had argued against this move, believing that future developments must be closely integrated with universities. Other voices in the UK prevailed however and in fact CERN has proved to be an outstanding scientific success and a model for international collaboration. Bubble chamber work at CERN required a new *modus operandi.* Most experiments were done by collaborations of several groups, often from different countries. Members of these groups required to spend time at CERN preparing the particle beam and running the experiment, the actual operation of the increasingly large and complex chambers being the responsibility of a resident crew. A share of the very large number of photographs was then brought back to the home laboratory for scanning, measurement and analysis. This latter work required teams of scanning and measuring technicians, sophisticated high precision machinery and, for the time, powerful computers.

The Glasgow laboratory was well supported by the Department of Scientific and Industrial Research (DSIR) in the development of scanning, measuring and computing facilities. The most important in a long succession of grants was in 1964 when Glasgow was the first UK university to be funded by DSIR to move into the second generation of powerful measuring and computing faculties with the installation of a set of 'scanning and measuring projectors', associated IBM 7040 and 650 computers and support for substantial operating staff. The subsequent success of this system owed much to Dr. C. J. Robinson, the senior staff member supported by the grant. The system was run day and night during the week with additional shifts at weekends and in due course celebrated its millionth measured event. Despite Dee's early scepticism concerning CERN he accepted the substantial growth and space demands of this work in the Department and the University found the necessary funds to modify the accommodation.

Using these facilities the Bubble Chamber Group carried out a long and successful series of experiments at CERN (and also at Rutherford Laboratory). The Group initially consisted of academic staff members I. Hughes, R. Turnbull and W. Morton and research students, later expanding with additional academic (R. Jennings, G. K. S. Browning, P. J. Negus, I. O. Skillicorn, J. M. Scarr), and grant supported staff. Research concentrated primarily on the interactions of K mesons but also used antiproton beams. These studies contributed substantially to the new field of particle spectroscopy which was one of the two pillars of the quark model now accepted as accounting for the basic structure of matter.

With the completion of the synchrotron programme in particle physics (it continued for a few years with a limited programme to provide synchrotron radiation beams largely used by visiting researchers, finally closing in 1973), a natural transition for some of the Glasgow physicists who had used the Glasgow machine was to the new higher energy (4GeV) electron synchrotron NINA. This machine was commissioned at the Daresbury Laboratory, a national laboratory for particle and nuclear physics opened in 1964 and intended to serve primarily universities in 'the north', complementing the Rutherford Laboratory not only geographically but also in facilities. J. C. Gunn, J. G. Rutherglen and J. M. Reid from Glasgow made important contributions to the launch of the new laboratory and the design of its accelerator. J. G. Rutherglen, G. Lewis, W. Hogg, K. Smith, R. Railton and a succession of graduate students, in collaboration with other UK groups, started work at NINA with a definitive measurement of the total γ-proton cross section over the energy range of the machine and then carried out a succession of increasingly sophisticated studies of the γ-nucleon interaction over the period 1965-1976.

The HT set shut down at the end of a highly successful programme of nuclear structure physics about 1960. With a view to a continuing programme in nuclear structure physics and with the end of the synchrotron programme in this area in sight it was decided in the early '60s to bid for funds for a new accelerator at Glasgow for nuclear structure studies. The choice of machine was a linear electron accelerator of energy 140 MeV. This choice filled a gap in the UK nuclear structure facilities, opening up a programme of nuclear studies using electrons and γ-rays. The bid was successful. The accelerator with its associated neutron flight paths was of such a size and required shielding such that there was no possibility of installation on Gilmorehill. Early studies aimed to build a new laboratory for this purpose on the Garscube Estate campus of the University, but there were problems with old mine workings, a suitable site could apparently not be found there and the Kelvin Laboratory housing the new accelerator with associated laboratories and workshops was built on government land in the grounds of the National Engineering Laboratory at East

Kilbride. This arrangement provided a suitable site at a nominal rent, close to the Scottish Universities Research Reactor laboratory, though across the city and 14 miles from the Department on Gilmorehill, a situation entailing much travelling by the staff involved and inevitably generating some fragmentation of the Department.

With the development of the new laboratory the Court made a new professorial appointment of George Bishop to the Kelvin Chair. Bishop came from the Orsay laboratory in Paris where he had been responsible for the electron accelerator, having earlier worked on electron-nuclear physics at Oxford. Other existing staff (J. Reid, K. Ledingham, J. Lynch, G. I. Crawford, A. McLeod) joined the Kelvin Laboratory Group which was strengthened by new appointments (S. Hall, R. Owens, D. Land, A. Johnson, J. Kellie, R. Singhal, A. Slight) over the following few years.

The Glasgow LINAC built by Vickers Research was designed by M. Kelliher who also supervised its construction and installation and later joined the staff of the Kelvin Laboratory as technical manager responsible for its maintenance and improvement. The machine got off to a good start on a programme of nuclear structure investigations in 1967. Bishop's particular field was the study of electron-nuclear scattering, then a relatively new technique for the study of the charge and magnetism distributions in nuclei and an area in which he had established a leading reputation. This work continued after Bishop's departure under the leadership of A. Johnson and A. Slight and collaborators. The photodisintegration of light nuclei became a major area of study by a group led by R. Owens and for some time this work benefited from a fruitful collaboration with a group at MIT. The last in a long series of experiments at Glasgow using a cloud chamber to study photonuclear processes was carried out at the LINAC by A. McLeod. The laboratory was designed to allow measurement of neutron energies by time of flight along long flight paths. Work by S. Hall and I. Crawford using time measurement to nanosecond precision made important contributions to the study of reactions using this technique. In another line of investigation J. Reid, in collaboration with Edinburgh physicists, studied the mass distributions of photofission fragments in an effort to understand the details of the disintegration process.

The Kelvin Laboratory also operated a smaller, 30Mev, linear accelerator which, as well as being used in the Glasgow programme, had a number of external (including commercial) users.

G. Bishop left Glasgow, initially on leave, in 1973 to go to the ISPRA Laboratory in Italy, resigning from the Chair in 1975. J. M. Reid, appointed to a Titular Chair in 1975, took over direction of the Kelvin Laboratory on Bishop's departure. Attempts to fill the Kelvin Chair with a nuclear structure physicist were unsuccessful and it remained vacant until 1989.

Another development of the late '60s was the start of work aimed at the detection of gravitational waves. Such radiation is predicted by Einstein's general theory of relativity, but the detection of these waves demands instrumentation of extraordinary sensitivity capable of detecting relative displacements now expected to be no more than 1 part in 10^{22}. R. Drever, who had specialised in measurements of great sensitivity, was encouraged to embark on the construction of a series of gravitational wave detectors by results reported by Weber in the USA who claimed to have seen gravitational wave signals. Weber's results are now generally believed to be spurious, but they initiated what has become a world-wide, multi-million-pound activity in the search for gravitational waves.

Initially the detectors consisted of sensors measuring small changes in the length of massive aluminium bars, but later devices depend on measuring the relative lengths of two orthogonal arms by interferometry of extraordinary sensitivity. The Glasgow group carried out much of the pioneering work in this field. Drever left Glasgow to take up an appointment at the California Institute of Technology in 1979 (continuing on a part time basis until 1989), but the Gravitational Wave Group continued to flourish under the leadership of J. Hough.

The early 1970s saw a number of important developments in the research profile of the Department. John Rutherglen and some others from the Daresbury group moved their research to CERN where they collaborated in experiments using a new photon beam of high energy to study the γ-proton interaction. While spending a sabbatical year working at CERN on this experiment Rutherglen was tragically killed in a motor accident in 1976. I. Skillicorn who had joined the Bubble Chamber Group from Stanford in 1970, and who had developed a collaborative project working at the DESY laboratory at Hamburg, assumed responsibility for the Glasgow γ-ray work at CERN. When this was complete he returned to develop a very fruitful programme of work on photon-photon interactions using the PETRA accelerator at Hamburg.

In the early 1970s it was becoming clear that important questions in particle physics required the study of quite complex processes but with a selectivity which the bubble chamber could not provide. Other techniques had been and were being developed however (spark chambers and then wire chambers) which did allow triggering on selected rare events and part of the Bubble Chamber Group led by R. M. Turnbull joined with a French group using a pioneering apparatus known as the OMEGA spectrometer at CERN. This was the first of a long and profitable series of such experiments. The bubble chamber group continued to work and indeed conducted its most ambitious experiment, a five million picture study of K^- meson-proton interactions, throughout the seventies, but the field of applications for bubble chambers was shrinking and the Group measured its last pictures in 1982 when its remaining members moved into electronic detector based techniques.

Theoretical physics

I have already described the establishment of the Theoretical Physics Chair with the appointment of J. C. Gunn in 1948. The Theory Group grew rapidly from the existing core (Sneddon, Guggenheimer, Touschek) by the appointment of new fellows and lecturers.

The 1960s decade was an age of academic expansion and the appointments made during that time under the guidance of J. C. Gunn set the frame for the theoretical research programme of the Department until the present day.

The majority of the Theoretical Physics Group led by R. G. Moorhouse have devoted their effort to work in particle physics (B. Bransden, A. Donnachie, D. Sutherland, A. T. Davies, I. Barbour, C. Froggatt, R. L. Crawford, C. T. H. Davies). In its early years the Group was internationally known for the quasi-experimental activity of the analysis of data to discover hadron resonances and to determine their properties. The Group has always had varied interests and emphasis has changed with development of the subject. Latterly work has concentrated on symmetries in particle physics and on lattice gauge theory; a method which tackles the formidable problem of evaluating the predictions of the 'Standard Quark Model' but which demands ever greater computer power.

Work on nuclear structure theory has been pursued throughout the period initially by N. McDonald and latterly by R. Whitehead and A. Watt. A particularly notable contribution to the field was the invention by Whitehead of a new and powerful method for nuclear shell model calculations. This method immensely extended the range of application of the shell model. It was exploited by Whitehead and Watt in the 1970s and has since become a very widely used technique.

Plasma physics has also been a long term interest in the theory department since the appointment of E. W. Laing in 1960, joined later by J. C. Taylor. The work has a close relation to experimental work on nuclear fusion and has involved continuing contact with the Culham Laboratory. This area of work has afforded a useful point of contact with theoretical work on stellar plasmas in the Department of Astronomy.

Some members of the Theory Group have worked primarily in other areas (G. Wyllie - condensed matter) and some have found their interests leading from one field to another - a transition easier for theoreticians than for experimentalists. Thus W. K. Burton moved from condensed matter to particle physics to questions in fundamental quantum theory and ultimately to the foundations of mathematics, N. MacDonald moved from nuclear structure theory to a variety of problems in mathematical physics including problems in biology and the theory of chaos while Whitehead has lately, *inter alia,* been applying sophisticated computer models in collaborative work with biologists.

Solid state physics

P. I. Dee retired in 1972 and the opportunity was taken to fill a serious gap in the physics repertoire at Glasgow. Many non-physicists do not appreciate that quantum theory has revolutionised condensed matter or solid state physics just as much as it has done for atomic, nuclear and particle physics, although the results of this revolution are apparent everywhere in the everyday world in the form of new materials, semiconductors and transistor based electronics. On the theoretical side G. Wyllie, who joined the Department as a Lecturer in 1950, worked on a variety of condensed matter problems throughout his career. Several attempts to establish an experimental research group in this area during Dee's tenure of the Natural Philosophy Chair proved unsuccessful however, due primarily to inadequate resources in a department with research concentrated on nuclear and particle physics. The lack of an experimental research activity in condensed matter undoubtedly also affected the capacity to give first class teaching in this area.

An experimental condensed matter group was established in 1973 with the appointment of R. P. Ferrier, a solid state physicist specialising in electron microscopy, to the Chair of Natural Philosophy, and shortly thereafter of J. Chapman and A. Long to lectureships. All three of these new members of staff came from the Cavendish Laboratory in Cambridge. The staff was later increased to five with the appointment of A. Craven to a Lectureship and P. Nicholson as a Research Technologist.

The research work of the Solid State Group has been primarily in electron microscopy, particularly in the study of micromagnetic structures, and in the study of the electrical, optical and structural properties of amorphous and latterly also crystalline semiconductors. The electron microscopy work started using a JEOL 100CX instrument purchased by the University. In 1978 the Group received a grant from SERC to purchase an HB5 Field Emission Scanning Transmission Microscope from VG Microscopes, while in 1991-92 the Group's pre-eminent place in the field of micromagnetic structural studies was recognised by SERC with the approval of a £1 million grant to purchase a new field emission electron microscope built to the Group's design by Philips of Eindhoven and optimised for the study of micromagnetic structures.

One of the strengths of the Group has been its contribution to the technical development of electron microscopy, in particular the development of differential phase contrast microscopy which allowed a major advance in quantitative studies on micromagnetism. The Group has also established many collaborations with industry and other universities.

Developments in the 1980s

In addition to the developments within the Solid State Group described above the 1980s saw substantial changes in the particle physics and nuclear physics programmes and the birth of a new research activity in atomic and molecular physics.

In the latter half of the 1970s it had become clear that many of the most interesting questions in particle physics required higher energy accelerators than were then available. The discovery of the theoretically predicted Z and W particles at CERN in 1983 confirmed the need for an electron-positron collider of sufficiently high energy to produce these particles in quantity. Such an accelerator, the large electron-positron collider LEP, was built at CERN, and provided its first colliding beams in the summer of 1989. High energy colliders operate with a small number of very large and sophisticated detectors normally designed, constructed and used by a large collaboration. A group of Glasgow physicists (I. Hughes, K. Smith. J. Lynch, C. Raine, A. S. Thompson joined later by I. Ten Have, W. T. Morton, J. M. Scarr and R. M. Turnbull) became part of a collaboration which built and used an elaborate detector known as ALEPH. The Glasgow Group made major contributions to the hardware by building a substantial part of the massive electron and photon detector in the Glasgow laboratory and by development work on the laser calibration system for the track detector. The ALEPH detector was ready to run when the first LEP operation took place in 1989, and, along with the other detectors at LEP, established unambiguously that the number of types of light neutrino in the Universe is three - a result with important consequences for particle physics and cosmology. ALEPH has since gone on to a highly profitable series of experiments testing features of the 'Standard Model' in particle physics.

Other members of the Experimental Particle Physics Group (I. O. Skillicorn, P. Bussey) who had been engaged in the programme at PETRA (DESY, Hamburg) joined a collaboration preparing a detector known as HI for experiments on the new DESY accelerator, a high energy electron-proton collider called HERA. This machine started operation in 1992 and experiments are now in progress.

I. S. Hughes took early retirement in 1989, and the Court agreed that in order to ensure continuation of the experimental particle physics programme with its substantial SERC support it should make an appointment to the Kelvin Chair in the field of experimental particle physics. D. Saxon from the Rutherford Appleton Laboratory was appointed to the Chair, taking responsibility for experimental particle physics and a major personal involvement with the HERA programme.

The Nuclear Structure Group also entered a new phase in the early '80s. The LINAC energy was too low to investigate the new problems of interest and after a period when it was used for a technical study relating to free electron lasers it was shut down in 1983 . R. Owens joined J. M. Reid as Associate Director of the Kelvin Laboratory in 1975 and it was Owens who led the Group into a programme using accelerators abroad when it developed a high performance facility for photonuclear experiments using a high quality beam from a new electron accelerator at the University of Mainz. This facility was used for a wide range of experiments. In particular it allowed the first quantitative study of short range effects in nuclei, the main unexplored area in nuclear structure. During 1987-91 the Mainz accelerator underwent a major upgrade to higher energy. The Glasgow Group was responsible for the design and construction of a new 'tagged beam' system which measures the photon energy for the interaction studied. The new facility has attracted some 15 other (mainly European) groups and has allowed the Glasgow Group to achieve a leading position in this field. Experiments to study components of nuclei other than protons and neutrons have begun and are beginning to yield interesting results.

During this period a new research activity has been established by K. W. D. Ledingham, joined later by R. Singhal and R. Jennings. The Group uses laser ionisation spectroscopy to detect and measure very tiny amounts of material. This technique can address a wide range of problems from pure physics problems in atomic and nuclear physics to very practical applied physics topics such as the detection of trace quantities of explosives and the study of impurities in semiconductor materials. It is interesting that this group was launched from the work done on laser ionisation of gases for the ALEPH calibration system.

In addition to the major research activities two smaller, but long term, medically related projects have been pursued in the Department. A. Mcleod has been responsible for the development of a computer controlled audiometer for the detection of defects of the ear. D. Land has been concerned with studies of the transmission and radiation of short wave electromagnetic radiation through and by body tissue and the development of instrumentation which uses such radiation to image internal features in the body in a totally non-invasive way. These instruments are now in use in hospitals for detection of tumours and for location and examination of inflamed tissue.

Students and the undergraduate curriculum in the post-war period

The update and revision of the curriculum following Dee's appointment were the start of an ongoing process. Although the basic Ordinary class curriculum would still seem quite familiar to Kelvin, students are now introduced to atomic physics and special relativity. In addition a second stream 'Ordinary General Physics' (now Physics B), 'Higher General Physics' and 'Advanced Ordinary General Physics' was introduced to provide classes with minimal mathematical demands for students not intending to go on to the Honours course. An extensive trial of 'Keller Plan' instruction involving only few lectures, detailed study guides and help session tutorials was run for nine years for part of the Ordinary class under the leadership of W. Hogg. The trial was a success in that the students liked it and keen students of even modest ability did well, while staff found the personal teaching satisfying. The scheme was however abandoned because of the very high demand on teaching time.

The small group supervision system (usually two-three students per staff member) introduced by Dee has not only survived, but has been extended to second year students. Honours class options have naturally changed and developed as the subject has evolved, while some cherished classical topics have had to be abandoned. An increasing number of students include a small research project in their laboratory work. The single Honours degree in Natural Philosophy has for some time been the most popular choice, but the long standing combined degree in Mathematics and Natural Philosophy continues to attract a steady stream of good students. Combined Honours with Astronomy, Electronic Engineering, Physics with Geology and Chemical Physics have generally smaller numbers.

Student numbers in Natural Philosophy rose rapidly between 1955 and 1965 (Figures 1 and 2). Over the past one or two decades however there has been a drop in numbers of students entering physics over much of the western world, and the UK has shared in this fall as has Glasgow natural philosophy, although to a lesser degree than many other UK departments. A gratifying feature has been that although first year numbers in natural philosophy have fallen Honours class numbers have not shown the same downward trend. The reasons for the decrease are complex, and the staff in Natural Philosophy have worked hard on contacts with schools to counter this effect. Special efforts have been made to interest more girls to take up the study of physics since, with a relatively small proportion of girls in physics classes, there is clearly a great reservoir of untapped talent. In latter years special 'Women in Physics' events have been held in the Department with a leading role being taken in this enterprise by Dr Norna Robertson. We have achieved some increase in the number of women students, but the effect so far has been small.

Merger with Astronomy and change of name

In 1986 the Department of Natural Philosophy merged with the Department of Astronomy to form the Department of Physics & Astronomy. This move was in response to pressures from the UGC, but in fact it recognised the very great overlap in interests and activity of the two departments both in teaching and research. Of the five members of staff in Astronomy at the time of the merger, four were theoreticians and there was an important common interest in the two departments in plasma theory.

Many staff in Natural Philosophy and many former students and staff regret the loss of the old name of the Department, both because of its long and distinguished tradition and because many of us believe that the name describes very well what we try to do. Nevertheless we must also recognise that much of the world at large does not understand what 'Natural Philosophy' means and that the old name does not convey the true image of a vigorous modern department.

Building refurbishment

A full and satisfying merger with Astronomy requires accommodation of both departments under one roof. At the same time the close-down of the LINAC at the Kelvin Laboratory at East Kilbride removes the need for the nuclear structure group to remain separated from the main Department. This latter move is rendered even more desirable by a very substantial increase in the rent for the Kelvin Laboratory following the change of management at the National Engineering Laboratory.

The existing Natural Philosophy Department cannot accommodate more staff and research activity without major modification. In addition much of the Department, particularly the old building, required basic refurbishment, rewiring and repair. The opportunity for this work arose in a UGC offer of funding for such projects early in 1988 and the University applied for and received funds for the work. The first move was to fill the mine workings under the old building where the pillars of coal supporting the old galleries were gradually crumbling away. A change which will sadden many for whom it holds fond memories is the loss of the large steeply raked lecture theatre which occupied a volume of space adequate for three floors. Some new and attractive laboratories are now emerging from the dust of the refurbishment, and hopefully the work will shortly yield a teaching and research laboratory well suited to its role in the 21st century.

Staff

One of the most significant changes in the Department has been in the size and composition of the staff. The academic and academic related staff numbers are shown in figures 1 and 2, and the contrast between the start and the end of the period is clear. Following the Second World War the staff increased sharply in two phases, the first following Dee's appointment and the second following the Robbins report in the second half of the 1960s. The surprising feature is that although the UGC-funded academic staff numbers then flattened out and have latterly decreased slightly, the total number of staff has continued to grow due to increasing numbers of grant supported research staff and research fellows. The days of dominance of a single professor as head of department have long since gone and such a system would be inappropriate to the large multifaceted Department of today. Of the four established Chairs in Physics only two - Natural Philosophy and Kelvin - are presently filled, but a further seven members of staff hold personal (Titular) Chairs. The Chair of Astronomy continues within the merged Department and is presently held by J. C. Brown. The position of Head of Department has not been tied to the Natural Philosophy Chair since the retirement of P. I. Dee in 1972 and has latterly been for fixed term periods (J. C. Gunn 1972-83, E. W. Laing 1983-86, I. S. Hughes 1986-89, R. P. Ferrier 1989).

The major expansion of staff numbers in the 1950s and 1960s which was so beneficial to the Department has inevitably brought its problems in later years, as the average age moves relentlessly upwards and it becomes very difficult to find posts for new young people. Nevertheless, the UGC 'New Blood' scheme and the desire of the University to maintain an active and successful research school have resulted in a number of important new appointments, while the steady flow of fixed term research assistant posts funded from external sources is an invaluable source of younger people.

Conclusion

In the hundred years since the institution of the Faculty of Science, the Department of Natural Philosophy has changed radically. From what was essentially a one man show it has developed into a college of over 70 professional physicists supported by technical staff, teaching and researching over all the major areas of modern physics. At the start of the period the University had as its Professor of Natural Philosophy the most distinguished physicist in the United Kingdom. Kelvin's reign was followed by almost 50 years when the Department was rather undistinguished in research though it fulfilled an important didactic role and produced many graduates some of whom went on to distinguished careers. In the favourable post-war climate Dee restored the Department to a leading position in the UK capable of teaching and research in a subject transformed by the developments of the first half of the century. The rejuvenated Department went on to make a world-class contribution to research in a variety of fields over the following 40 years. As we approach the end of the 20th century physics remains as interesting and challenging as ever but the financial, demographic and bureaucratic problems faced by staff are severe. You have to be an optimist in order to research successfully in physics (and probably in any field of science) and on the basis of the current strength of the Department in its staff and external support I am optimistic for the future of physics in Glasgow in the 21st century.

Acknowledgements
The opinions expressed in this essay are those of the writer, but I am much indebted to the following for information and reminiscences: Ian Andrews, Jimmy Atkinson, Professor Sir John Gunn, Dr G. M. Lewis, Seumas McNeill, Dr D. Stewart, Dr G. Wyllie, Professor J. M. Reid.

I am also grateful for information received from several present members of the staff of the Department of Physics & Astronomy: Dr J. Cumming, Professor R. Ferrier, Professor J. Hough, Professor G. Moorhouse, Dr P. Negus, Professor R. Owens, Professor K. Smith and Mr G. Baird.

My researches in the archives were guided by the archivist Mr M. Moss, for whose advice I am particularly grateful. A source of data for the earlier part of the period which I found useful is the series of *Financial and Statistical Reports to Parliament* by the four older Scottish universities. I also recommend the University *Calendar*, particularly in earlier times, as a surprisingly rich source of information.

References

1. H. Stein: After the Baltimore Lectures . . . published in Kelvin's *Baltimore Lectures and Modern Theoretical Physics* ed. Karagon and Achinstein, MIT Press, 1987.
2. Letter T. Alty to H. Hetherington, 1943
3. L. Badash: Ernest Rutherford and Theoretical Physics, published in Kelvin's Baltimore Lectures . . . op. cit.

Ian S. Hughes is Emeritus Professor in the Department of Physics & Astronomy at the University of Glasgow.

Botany

Arthur D. Boney

The teaching of botany in the University dates from 1704 when a small physic garden was established in the grounds of the Old College in the High Street and John Marshall, 'Chirurgeon and Bottanist', was appointed its overseer and the first teacher of botany. Until his death in September 1719 he supervised the running of the garden and instructed students in the recognition of plants used as sources of 'simples' or drugs. After Marshall's death a joint professorship of Botany and Anatomy was created by Royal Warrant and Thomas Brisbane appointed the first occupant of the Chair in 1720. The professorship remained a joint one with a succession of occupants until 1818 when a separate Chair of Botany was founded by George III. The physic garden gradually fell into disuse after the early 1800s, mainly due to the cumulative effects of toxic fumes from a nearby type-foundry in the grounds, and until 1820 the subject was taught without benefit of such a garden, and with plant supplies purchased each year from local gardeners. William Jackson Hooker, Regius Professor 1820-1841, gave his lectures in the Sandyford Garden of the Royal Botanic Institution, an eight acre garden lying between Argyle Street and Sauchiehall Street. The University had three directorships on the governing body of the garden due to financial support given on its foundation in 1818. The Professor of Botany was always one of these directors. With the removal of this Botanic garden in 1842 to its present-day site, the botany lectures were again given in the Old College by successive professors because of the lack of a lecture room in the buildings of the botanic garden, so that student visits to the more distant garden would have been restricted somewhat. After 1870 and the removal of the University to the new buildings at Gilmorehill, the botanic garden now became more available for students, but the lectures were given in a lecture room in the north-west tower of the East Quadrangle at Gilmorehill. From its beginnings in 1704 the teaching of botany remained a province of the Faculty of Medicine, and until the late 1880s the sole annual teaching requirement of the professor was some 40 to 50 lectures to medical students in the summer session.

Until late in the 19th century the main emphasis of the botany lectures in the University was on floristic and systematic aspects, in common with the general trends of botanical studies in Britain. Meanwhile fundamental advances in the subject were taking place on the continent, especially in Germany. In the later years of the last century aspiring young British botanists found it essential to embark on periods of study in German universities in order to keep pace with current advances in the science. However, by the time that the Faculty of Science was established in 1893, a significant and fundamental change had taken place in the teaching of botany in the University, a change which came about after the appointment of 26-year old Isaac Bayley Balfour to the Regius Chair in 1879. Balfour came to Glasgow fresh from studying on the continent, and in his first year 'bartered' the lecture room in the north-west tower for two rooms on the ground floor in the East Quadrangle, the larger one to serve as a teaching laboratory and the smaller one opening off it for a preparation room, and for himself. For his lectures he borrowed the lecture room of a professor whose lectures were confined to the winter session. The University agreed to install a water supply and provide furnishings for this laboratory, whilst Balfour himself purchased the microscopes and other teaching equipment. Under the 'farmed chair' principle then in practice Balfour collected the student's fees as part of his own salary and that of any assistant he employed. Balfour's long-term plan was to establish a botanical institute in the University, but situated in the Botanic Garden. As with his predecessors, he was a Director of the Botanic Garden. At one stage he had hoped to acquire North Park House just across the road from the Garden as his institute, but this plan fell through. (North Park House was to become Queen Margaret College for Women Students in 1884; today, much enlarged, it is the headquarters of BBC Scotland). Balfour succeeded in getting the main range of plant houses rebuilt, and established the Kibble Palace as a winter garden, so rescuing it from military band performances and other entertainments, including performing dogs. Balfour left Glasgow in 1885 on his appointment to the Sherardian Chair of Botany at Oxford, his ambitious plans unrealised, but having rejuvenated the subject in the University. He was succeeded by Frederick Orpen Bower.

Yorkshire-born F. O. Bower had graduated in 1877 as a member of Trinity College, Cambridge. In his final year he had been taught by S. H. Vines, a newly appointed Lecturer who brought to Cambridge the 'New Botany' which had emerged from the 'Botanical Renaissance' in Britain in the early 1870s. The Normal School of Science at South Kensington was the principal source of the 'Renaissance' in Britain. The Normal School had been formed in 1871 by an administrative metamorphosis of the Royal School of Mines, and its principal function was to train science teachers. T. H. Huxley was its head with the title of

Dean (the latter a source of amusement to Huxley and his friends). From 1875 W. T. Thiselton-Dyer, later to be Director of Kew Gardens, was responsible for botany at the school, and his courses were run on similar lines to those in many continental universities, with laboratory classes as the basis for practical work, and greater emphasis on the study therein of structures, life histories, and simple physiological observations, with the purpose of balancing the overemphasis on systematic work. Over the years Thiselton-Dyer ensured the spread of the new 'gospel' by employing aspiring young botany graduates as demonstrators, of whom Vines had been one. For Bower this glimpse of the 'New Botany' was a Damascene-conversion experience, in common with other young botanists in Britain. For two years he had suffered under a moribund botany course at Cambridge with its pronounced floristic bias, and taught by the Professor, Charles Cardale Babington, then in his late sixties, and not an inspiring lecturer. In the spring of 1877 Bower studied for a short period under Julius von Sachs at Würzburg, and then spent the 1879-80 session in the laboratory of Anton de Bary at Strasbourg. His mentors were acknowledged master botanists of their time, and each was to have a profound influence on Bower's later scientific outlook. After employment in a number of Assistantships, Bower was appointed Lecturer in Botany at the Normal School in 1881. Here he was able to develop his own teaching methods in the 'New Botany'. There was also time in the academic year for him to pursue research at the Jodrell Laboratory in Kew Gardens. He was now in an ideal situation to gain worthwhile experience and await promotional opportunities.

The first promotional opportunity came in 1884 with the vacancy of the Sherardian Chair at Oxford. As he was to learn soon afterwards, his application came second to Bayley Balfour in the opinion of the electors. The resulting vacancy at Glasgow he ignored - it was of no interest to him in comparison with his position in London, with a view to more attractive chairs turning up in the near future. But in May 1885 he was to find himself appointed to the Glasgow Chair, albeit unwillingly. Since he was to serve the University for the next 40 years some enquiry into the background to his appointment seems well justified. His enjoyable academic life in London came to an abrupt end in early March 1885. He was interrupted whilst working in the Jodrell Laboratory by the entry of Sir Joseph Hooker, Director at Kew, Thiselton-Dyer, then the Assistant Director, Professor Daniel Oliver of University College London and J. G. Baker, the four chief officials at Kew. In reply to Sir Joseph's enquiry Bower stated that he had no intention of applying for the Glasgow Chair. He was duly informed

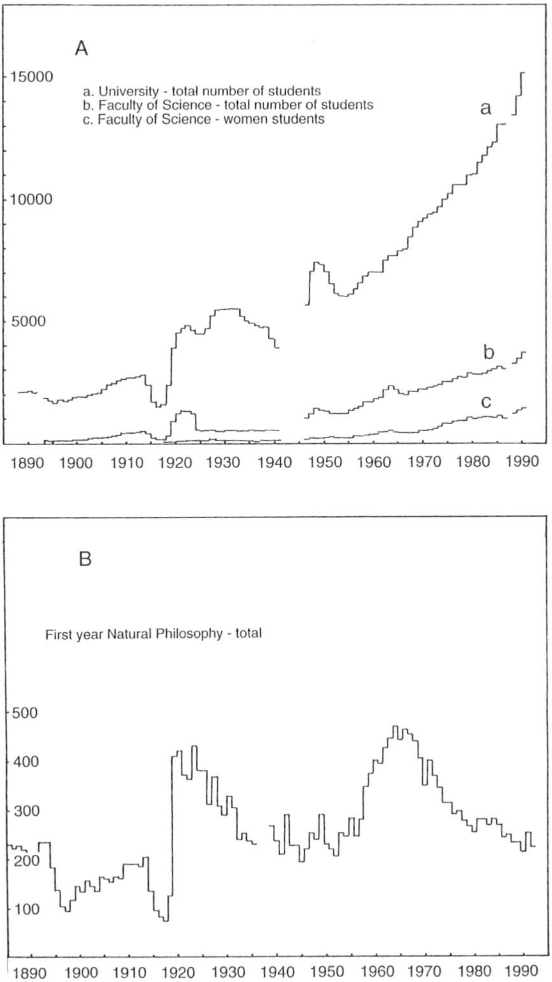

Figure 1

Natural Philosophy numbers 1890 - 1990

A: (a) University: total number of students
(b) Faculty of Science: total number of students
(c) Faculty of Science: women students

B: First year Natural Philosophy: total number of students

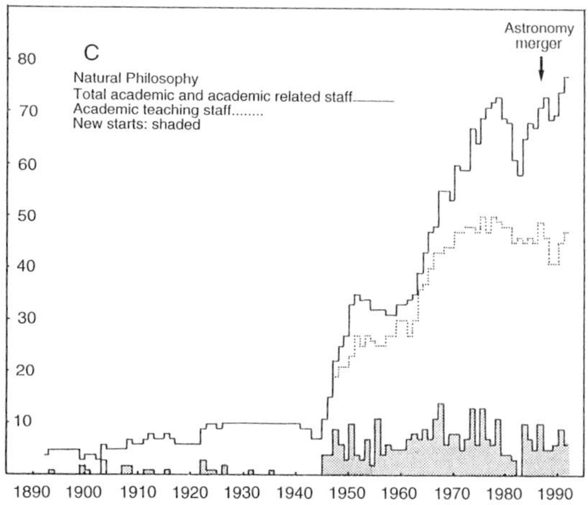

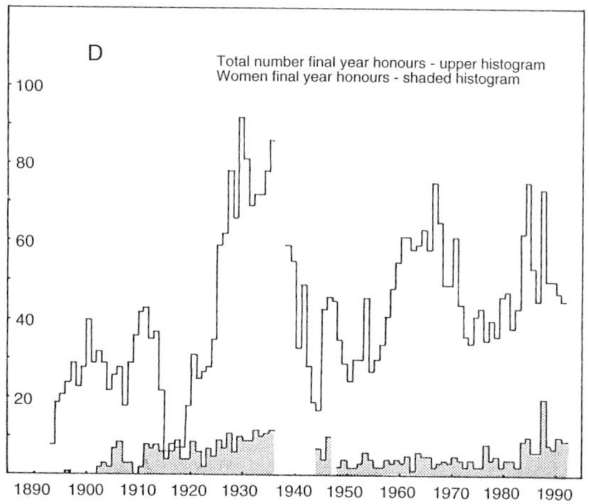

Figure 2

Natural Philosophy numbers 1890 - 1990

C: Natural Philosophy: Full line - total academic and academic related staff; dotted histogram - academic teaching staff; shaded histogram - new starts. Note that the difference between the full line and dotted histograms is short term grant supported staff and fellows. New starts are mostly fixed term research assistants and fellows.

D: Final year honours students in the Natural Philosophy Department: upper histogram - total number, shaded histogram - women students. Gaps in histograms correspond to missing data.

Figure 3

Lord Kelvin with compass.

Figure 4

P. I. Dee with staff and the 1971 final year outside the door of the Old Natural Philosophy Building.
Front row right to left: G. Wyllie, R. Drever, A. Cockroft, E. Laing, J. Gunn, P. I. Dee, J. Rutherglen, W. McFarlane, I. Hughes, R. Jennings, J. Taylor.

Figure 5

Clockwise from top:

Large lecture theatre in the Old Natural Philosophy Building.

Front entry of the New Natural Philosophy Building.

330MeV electron synchrotron from above (W. McFarlane and technician at right).

Figure 6

Figure 7

Figure 8

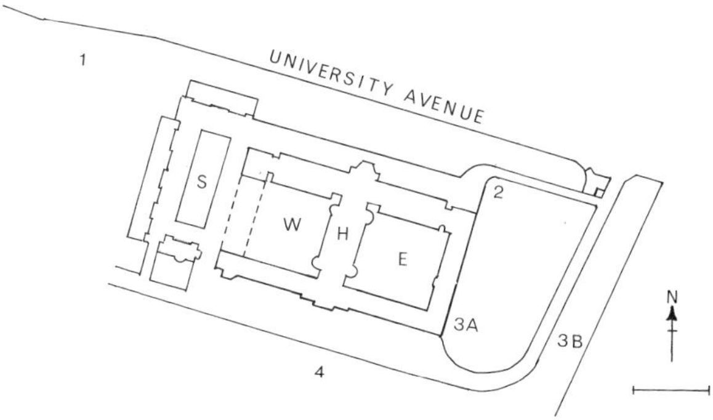

The illustration above shows the four sites which were under discussion for the joint Botany and Engineering Buildings, 1894-1898.

1.	*The tennis courts site, where the Botany Building now stands.*
2.	*The site on ground behind Pearce Lodge.*
3.	*The two sites beyond the East Quadrangle which were the cause of much argument; 3A is about where the present Engineering Buildings now stands.*
4.	*The site on the south front above Kelvingrove Park.*
S.	*The Square.*
W.	*West Quadrangle.*
H.	*Bute Hall.*
E.	*East Quadrangle.*

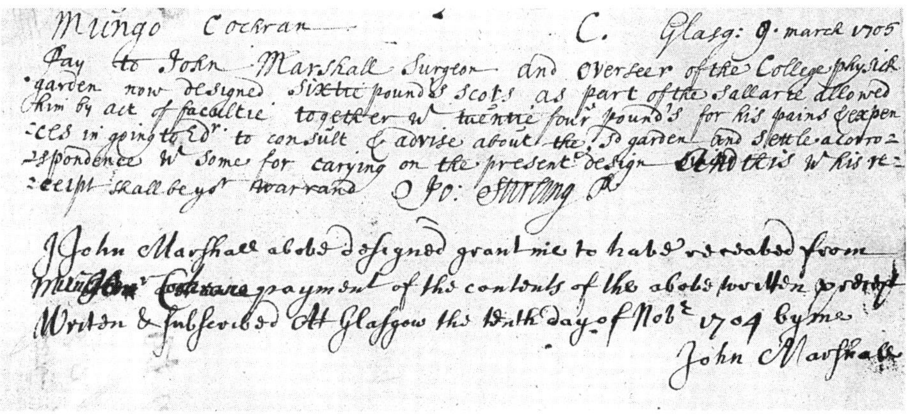

Mungo Cochran C. Glasg: 9 march 1705
Pay to John Marshall Surgeon and Overseer of the College physick garden now designed Sixtie pounds Scots as part of the sallarie allowed him by act of facultie together wt twentie four pounds for his pains & expences in going to Edr to consult & advise about the sd garden and settle a corre=spondence wt some for carying on the present design And this wt his re=ceipt shall be yor warrand Jo: Stirling R

I John Marshall above designed grant me to have received from Mungo Cochrane payment of the contents of the above written precept Writen & subscribed At Glasgow the Ninth day of Nobr 1704 by me
John Marshall

Figure 9

One of the few tangible links with John Marshall of Botany; a precept dated 5 March 1705 for part of his salary and some expenses in going to Edinburgh for consultations.

Figure 10

The opening of the Botany Building in June 1901.
The official party posed in the Elementary Laboratory.

1. G. Farlow (Harvard)
2. K. Timiriazeff (Moscow)
3. I. Bayley Balfour (Edinburgh)
4. Lord Lister
5. Sir Joseph Hooker
6. Principal Story
7. J. M. H. Traill (Aberdeen)
8. Lord Kelvin
9. D. H. Scott (Jodrell Laboratory, Kew)
10. F. O. Bower
11. W. Thiselton-Dyer

Figure 11

The Zoology building as it was in the 1920s. The teaching laboratory (last three window bays on the right) rose the full height of the building to capture all available light - an objective defeated by the construction of the Chemistry building alongside it in the 1930s.

Figure 12

Graham Kerr's Zoology Museum with daylight entering from above. Its splendid interior was at one time used for University banquets such as the rectorial luncheon for Lord Birkenhead whose controversial speech ("The world continues to offer glittering prizes to those who have stout hearts and sharp swords") went down in history. The space beyond the Doric columns has now been comandeered for other purposes and the construction of roof laboratories above the Museum has cut out the daylight.

Geology portraits

Figure 13:

Top left:
Professor H. D. Rogers

Top right:
Professor John Young

Bottom left:
Dr John Young

Bottom right:
Professor J. W. Gregory

Figure 13

Figure 14:

Top left:
Dr G. W. Tyrrell with, on the right, Mrs Tyrrell and, on the left, Archie Lamont

Top right:
Professor E. B. Bailey

Bottom left:
Professor A. E. Trueman

Bottom right:
Professor T. N. George

Figure 14

that there had been a 'hitch' in the appointment, and the four of them considered it his duty to apply. Bower knew that he was 'bound to obey', and his agreement set in train a whirlwind series of events, including the supply of numerous printed testimonials, the preparation of his Royal Commission by the Home Office, an interchange of correspondence with Principal John Caird at Glasgow, and finally on May 12, 1885, the reading before the Senate of his Latin essay ('De Somno Plantarum' - 'On the sleep of plants'), which, being duly 'sustained,' and after some statutory declarations, he was formally admitted to the Chair. Meanwhile he had already begun (as Mr Bower!) his course of lectures to some 200 medical students.

The question of the 'hitch' which brought Bower unexpectedly to Glasgow remains largely unanswered. In his 1938 autobiographical sketch he stated that the underlying reasons were unknown to him, but he was ever a discreet personage. The Home Office destroyed the relevant papers in June 1885. Thiselton-Dyer, deeply involved in the affair, told Bower in November 1912 that he had burned the relevant file at Kew soon after confirmation of the appointment. From what can be surmised, the 'hitch' had political overtones. At the time Gladstone was in the process of finalising his Home Rule for Ireland policy. In Dublin there was a troublesome and unpopular Professor of Botany. He was William McNab, an Edinburgh M.D., who in 1879 tried to add the Directorship of the Glasnevin Botanic Garden to his professorship at the Royal College of Science. The successful appointee at Glasnevin, Frederick Moore, was known to Thiselton-Dyer from the latter's time in Dublin in the early 1870s as Professor of Botany at the Royal College of Science, and predecessor of McNab. The unsuccessful McNab was given the conciliatory and powerless title of Scientific Director of the Botanic Garden, but remained a source of trouble. Thiselton-Dyer would have known of McNab's machinations. To remove an unpopular Scot from Dublin and replace him with someone more amenable would have brought some political kudos, however small, to Gladstone, and such a move may well have been put to the Glasgow Senate. Indeed, McNab seems to have assumed that the appointment was a cut and dried affair. Principal Caird would have consulted Sir Joseph Hooker as a senior botanist in Britain, an M.D. of the University, and the son of a distinguished Regius Professor of Botany. Sir Joseph would have consulted Thiselton-Dyer, and thereafter there was a concerted move to prevent McNab's appointment at Glasgow, with the result that none of the candidates were considered suitable. Hence the subsequent haste to place Bower in the Chair.

Somehow Bower struggled through that first course of lectures and laboratory classes in 1885, in what was to him an alien academic environment, and with students the like of whom he had never before experienced. Even under these circumstances however, he made his mark, as seen in letters he received in later years from one-time medical students of that first summer session. To his complete surprise he then learned that the summer programme was all that was required of him. In consequence he spent the following late autumn and winter months botanising in Ceylon, returning 'much refreshed' for the 1886 summer session. Whilst coping with his teaching - he was to continue lecturing to first-year medical students for the remainder of his time at Glasgow, regarding each lecture as a 'sporting event' - he began to find the botanical isolation unbearable. There was no-one with whom he could discuss his developing research interests, and he spent as much time as possible away from Glasgow visiting other centres, including the Jodrell Laboratory. This botanical isolation brought him to the brink of resignation after two years, but such a move was forbidden by Thiselton-Dyer, who reminded Bower that it was his duty to build up a flourishing department at Glasgow. Then in 1888 Bayley Balfour returned to Scotland on his appointment to the Edinburgh Chair, combined with the Keepership of the Royal Botanic Garden. Bower now had a close neighbour who was botanically a kindred spirit, and a close friendship grew between the two. Together they dominated botany in Scotland for some 32 years, and were quite prepared to exert their joint influence south of the border if occasion demanded - and often did so.

The institution of the Faculty of Science in 1893 found Bower deeply involved in a major campaign on behalf of his Department. His accommodation remained as in 1885, two rooms and an attic herbarium, and a borrowed lecture room - a situation which he later described as "a thing of threads and patches". Almost from the outset he had sought to complete Balfour's plan of a botanical institute in the Botanic Garden. For both Bower and Balfour a lecture room and some systematic flower beds were insufficient; in addition they wanted a building with adequate laboratory accommodation for teaching and research, a properly organised herbarium, a museum and a library. There could hardly have been a more inopportune time for such an undertaking. By the late 1880s the Royal Botanic Institution which still owned the Botanic Garden was in severe financial straits, heavily in debt to the Glasgow City Council, and the very survival of the Garden was at stake. Instead of pressing on with building plans Bower found himself fighting to save the Garden, and to prevent the sale of valuable plant collections. In 1888, with the likelihood of an institute in the Garden seemingly remote, he proposed to the Senate that suitable laboratory space for Botany could be achieved if some attic rooms in the East Quadrangle were enlarged by raising the roof area, and with Botany returning to its lecture room in the north west tower the whole arrangement could be *en suite.* It took the Senate three years to consider the proposal, and nothing then came of it. Meanwhile the Botanic Garden had been taken over by the City Council, and whilst Bower may have had some lingering hopes of still establishing his botanical institute within its bounds, these hopes were dashed in February 1894, when he was informed that the Council could make no guarantee that the Garden would remain on its site for the next 20 years, and in consequence the University could not consider any financial undertaking regarding Botany requirements therein. The Senate then proposed that the new Botany building should now be erected within the grounds at Gilmorehill. As is well known, the Botanic Garden survived, and close links remain with the Botany Department. From a long standing agreement the Garden serves as a source of plant material for both teaching and research.

A measure of Bower's desperation in trying to establish a department commensurate with the broadening base of the botanical sciences is to be seen in a letter he wrote to the University Court on 10 May 1893 - about four weeks prior to the institution of the Faculty of Science on 5 June. His letter was in response to the Court's reply to an earlier letter in which he had stressed the now dire need for improved accommodation, to which the Court had offered a temporary relief. His reply was to the point:

"No temporary arrangement can be satisfactory when things are as bad as they are - no class room, no sufficient laboratory, no botanical museum or proper Herbarium. All this must be sooner or later provided or else Glasgow University will have to take the hindmost place as regards Botany: in which case it will be hardly worth my while to remain here."

Eight years on from his appointment he was no longer a somewhat overawed junior professor. He was now an Sc.D. of Cambridge, and had been elected F.R.S.E. and F.R.S. Despite the attendant difficulties of accommodation he was by now well on the way to becoming one of the foremost research workers in his field. Perhaps the clear threat in the letter and a belated recognition of the strength of the case at last stirred up a positive response. Within less than 12 months the Court was in consultation with architects regarding new buildings for both Botany and Engineering. Archibald Barr, Professor of Civil Engineering 1889-1913, was equally ambitious in his requirements for adequate accommodation. The initial plans were for the two buildings to be erected close together, and four sites came under consideration. Tennis courts then occupied the ground on which the Botany building now stands, but as a site for the two buildings it was considered unsuitable. The second site was on the ground behind Pearce Lodge, and the third on ground beyond the East Quadrangle, where the Engineering Departments now stand. A fourth site examined was on the ground above Kelvingrove Park and below the main front of the University. The third site became the main focus of attention and planning, and this met with the concerted opposition of the medical professors, some of whose departments occupied cramped quarters in the East Quadrangle, with overflow classes in temporary iron huts on the neighbouring ground. Their opposition was mainly aimed at Barr, whose ambitions they suspected would not rest with one building for his Department. Arguments, counter-arguments and planning discussions went on for four years, giving rise to some heated discussions in Senate. The settlement came with the intervention of Lord Kelvin in March 1898. Dismayed by so much discord in Senate he held discussions behind the scenes and proposed that the Engineering building should be built on the east side beyond the East Quadrangle, and that the Botany building be erected on the tennis court site. It is a measure of Kelvin's standing that these proposals were almost immediately adopted. The Botany building cost some £19,000, with funding from the Charles Randolph Trust and the Bellahouston Trustees. The building was opened by Sir Joseph Hooker on 13 June 1901 as part of the University's ninth jubilee celebrations. After 16 years Bower at last had a purpose built institute, with adequate space for two large teaching laboratories, an herbarium on two floors with a small library, a museum occupying three floors, a lecture room for some 300 students, rooms for himself and staff members, and a workshop. Nevertheless, some of his professorial colleagues considered the accommodation altogether too spacious.

In 1893 the teaching requirements of the Department had increased from the 40-50 summer lectures to medical students. Botany was now offered as a subject for both the M.A. and B.Sc degrees, with courses spread over the academic year. Lectures were also given at Queen Margaret College. The 'farmed chair' process ceased after 1889 with now a centralised control of finances, and Bower was able to appoint assistants (most of whom became lecturers in due course). The roll-call of his assistants and lecturers over the years includes a number who went on to become botanists of distinction:

• J. G. Willis, Assistant 1893-4; later Director of the Peradinaya Botanic Garden in Ceylon, and then Director of the Rio de Janeiro Botanic Garden. Elected F.R.S. in 1921.

• W. H. Lang, Assistant and Lecturer 1894-1909; first Barker Professor of Cryptogamic Botany, University of Manchester, 1909-1940. Elected F.R.S. in 1911.

• D. T. Gwynne-Vaughan, Assistant and Lecturer 1896-1907; Head of Botany at Birkbeck College, London 1907-1909; Professor of Botany, Queen's University Belfast, 1909-1914; Professor of Botany, University College, Reading, 1914-1915; died of T.B. of the lung in September 1915, aged 44 years.

• A. A. Lawson, Lecturer 1907-1912; Professor of Botany, University of Sydney, 1912-1927.

• J. M. F. Drummond, Lecturer 1909-1921; Director, Scottish Plant Breeding Station, 1921-1925; Regius Professor of Botany, University of Glasgow 1925-1930; Professor of Botany, University of Manchester, 1930-1957.

• J. McLean Thompson, Assistant and Lecturer 1911-1921; Professor of Botany, University of Liverpool, 1921-1952.

• C. W. Wardlaw, Assistant and Lecturer 1921-1926; Barker Professor of Cryptogamic Botany, University of Manchester 1940-1958; Professor of Botany, University of Manchester 1958-1964.

It was the association of Bower, Lang and Gwynne-Vaughan - the Glasgow 'triumvirate' - which firmly set Botany in place as a subject for an Honours degree, and established the Department's high standing as a centre for research.

The B.Sc. degree had been instituted by Ordinance in 1872, and included botany as one of the subjects in the biological sciences group, to be taken with three other subjects for a degree which in the 1890s became one in Natural Sciences. Three main aspects were covered - morphology, physiology and systematic botany. The physiological component was limited to carbon assimilation ('amylogenesis'), respiration and water uptake and movements. There seems to have been little demand for any botanical participation before the 1890s. The subsequent developments by Bower and his associates which led on to the Honours degree appreciably broadened the scope of the subject, covering the external morphology, anatomy and physiology of plants, features of their geological history and a general knowledge of the leading types of the main plant groups. The 'type method' was very much in the South Kensington tradition, entailing (as Bower explained each year in his first lecture to elementary students) - "a detailed study of a few forms (which) was better than a superficial study of many - the type being the plant which represents most fully the largest number of important characteristics of the group to which it belongs." Here, with the accompanying laboratory classes, was a full flowering of the 'New Botany' of Bower's limited undergraduate experience and his teaching at the Normal School. This 'type method' approach continued down through the years, although with some necessary modifications - some of the 'types' of the 1890s have since been found atypical of their group. Plant physiology at Glasgow did not become a prominent component of the curriculum until the appointment of J. M. F. Drummond in 1909, with lectures on metabolism, growth and irritability. Within these broader outlines, however, the advanced teaching at Honours level was of an informal nature, with discussions on recent advances linked with current researches - forerunners of the 'option' or 'module' approaches of later years. The standard of teaching in the Department was a high one, with Bower very much a leading exponent, and a standard maintained by his shrewd choice of colleagues. A strict disciplinarian, sometimes known to exhibit a fiery temper, Bower was well able to cope with large elementary classes. At the same time he fostered a deep respect from generations of students. Similarly, all laboratory classes were meticulously planned and executed. Day-long field excursions during the summer months were of regular annual occurrence, frequently to the nearby highlands to study alpine floras.

From 1890 until 1925 the Botany Department became internationally recognised as a centre of excellence for research on plant morphology. From Bower's time at the Jodrell Laboratory in the 1880s until after his retirement, all aspects of his researches were a constant source of pleasure. In broad definition his researches are embodied in the general title of comparative morphology and anatomy in relation to phylogenetic relationships, concentrating on the archegoniate plants. It was comparative anatomy and morphology under the Darwinian influence expressed in the *Doctrine of Descent.* Two phases in his research can be identified. Phase one, from about 1890-1908, culminated in his book, *Origin of a Land Flora,* in which he postulated for lower plants the pathways by which the sporophyte generation became the dominant terrestrial form, in support of the antithetic theory of alternation of generations. At the time the book was regarded as the major contribution to morphological theory, and it stimulated much discussion both for and against the hypothesis therein. He came to modify his views in 1935 with *Primitive Land Plants.* In the second phase he concentrated on a study of the ferns in relation to phylogeny, work which resulted in a series of major papers and three books, the latter published between 1923 and 1928. As a broad generalisation he concluded that in ferns there seems to have been evolutionary change leading to greater adaptation to life on land. During the 1920s his interests centred on aspects of size and form in vascular plants with regard to stelar structures, and which resulted in his book *Size and Form in Plants,* also published in 1930, but which received some critical reviews over which he was not too pleased. His colleagues were similarly active in research. Lang worked on liverworts and on apospory in ferns and Gwynne-Vaughan on fern stelar anatomy. Lawson's research was on the Gymnosperms; McLean Thompson worked on floral morphology and Wardlaw on fern stelar anatomy. In 1904 an important association developed between the Glasgow Department and Robert Kidston of Stirling, a distinguished palaeobotanist. The main collaboration was firstly between Kidston and Gwynne-Vaughan until the latter's death in 1915. Together they produced a classic series of papers on the fossil Osmundaceae. After Gwynne-Vaughan's death, Lang joined Kidston and their researches on the Rhynie Chert fossils were similarly to be regarded as classic, and to this day are considered amongst the most outstanding contributions to knowledge of early vascular plants. Although this significant research was carried out after the members of the 'triumvirate' had gone their separate ways, it was to the Department that Kidston, who died in 1924, bequeathed his unique collection of memoirs, books, and slides of fossil plants, including many type specimens.

The great success of the Department as a centre for research on morphology and anatomy was in part due to the nature of the practical work involved. Bench space and apparatus requirements were relatively small. A good microscope was of course essential, together with skills at cutting sections (usually by hand; Bower rarely used a microtome) and the staining and mounting of sections. An ability to draw accurately was of prime importance. There was the ideal combination of a fruitful field of research waiting to be carried out and a group well able to exploit the opportunity. Bower gave formal notice of his impending retirement in May 1924 and in his 69th year. His great hope was the W. H. Lang would succeed him. Lang gave the move serious consideration, but before he made any application Manchester made some quick moves to retain him, both financial and making the Barker Chair a research position. H. H. Dixon, Professor of Botany at Trinity College Dublin, a distinguished plant physiologist, was appointed to the Glasgow Chair, but soon afterwards withdrew whilst professor-designate. His withdrawal was over some difficulties regarding pension rights, and there seem to have been misunderstandings on both sides. At the same time Trinity College made moves to retain him. Bower was naturally distressed by the unforeseen turn of events. His letter to Dixon, whilst expressing understanding of the pension problem, also included part of a sentence which shows that he knew more about the political background to his 1885 appointment than he was prepared to admit later in his 1938 autobiographical sketch, namely, in comparing 1884-85 and 1924-25, there was "trouble in both cases from Dublin".

John Montague Frank Drummond was appointed Bower's successor in 1925, the latter continuing in post for the interim period. Drummond, a Cambridge graduate, had come to Glasgow in 1909 from Armstrong College, Newcastle-upon-Tyne, as Lecturer in plant physiology. His academic career was interrupted in 1914 when he volunteered for military service and he was commissioned in the Highland Light Infantry in 1915. He was to see active service both in the Middle East and in France, and was demobilised with the rank of captain in April 1919. During his time at Glasgow he had studied the structure and functions of hydathodes, and investigated the fungal infections of lens adhesives in range-finders manufactured by Barr and Stroud. In 1914, he published a translation of Haberlandt's *Physiological Anatomy of Plants.* When not involved in battles with the Turkish Army he botanised extensively, building up a collection of Middle East plants on which he worked with the staff of the British Museum (Natural History) after his demobilisation. His service from 1921 as Director of the Scottish Plant Breeding Station at Corstorphine in Edinburgh led to an appreciation of the importance of genetics, plant pathology and plant ecology and geography in undergraduate courses, and these topics became part of the teaching programmes after his appointment. Notably these were topics (particularly ecology) excluded from the curricula in Bower's time. In his letter to the University Court after retirement, Bower had stressed the need for increased accommodation for plant physiology research in the Department, and for a Lectureship in Mycology (he used the less euphonious title of 'fungology'). However, the Lectureship was not established until 1928, and no additional laboratory space became available in Drummond's time.

Drummond came to Glasgow with the intention of maintaining the strong morphological tradition, and to break down the barriers between the academic and applied aspects of the subject. Whilst agricultural botany had been offered as a component of the B.Sc. (Agriculture) degree under an 1889 Ordinance it was never a strong side in Bower's time. It was under Drummond that agricultural botany became a principal subject for an Honours degree in Agriculture under the 1924 Ordinance. The B.Sc. Honours in Botany continued on similar lines as in Bower's time but with the additions already mentioned. For the reasons already given, Drummond lacked a strong research background although he was a sound teacher. Whilst departmental research was encouraged it remained principally in the morphological sphere. Two of his members of staff had been Bower's appointees, and a Lecturer in Mycology was the third staff member appointed in 1928. Three Assistants were in post in 1928, two of whom, G. Bond and D.G. Catcheside, had been appointed by Drummond, and were in later years to be elected Fellows of the Royal Society. Drummond's wartime experiences, in which he had emerged unscathed from both the fighting and after escaping from a torpedoed troopship, left him with a lasting affection for the Army, and he was Commanding Officer of the University O.T.C. throughout his professorship. He held strong views on one aspect of departmental organisation. In January 1925, with his appointment assured, he informed Bower that he had been approached by J. W. Gregory, Professor of Geology, "to restore that horrid chimeraea (sic) 'Biology'". He assured Bower that he would dissent most strongly from any such proposition. With Drummond's appointment to the Chair of Botany at Manchester in 1930 something of an exchange took place, with John Walton coming from Manchester to the Glasgow Chair.

John Walton, a Cambridge graduate, had studied there under Professor A. C. Seward and Dr H. Hamshaw Thomas, both distinguished palaeobotanists. Seward, who was to be knighted in 1936, was expert in both botany and geology and was an excellent teacher. Hamshaw Thomas was especially skilled in teaching the techniques involved in the examination of compression type plant fossils. Walton's training came towards the end of what he himself described as the "grand period of British palaeobotany, from 1880-1920". At Manchester in the early 1920s, however, he joined a Department in which the two professors, F. T. Weiss and W. H. Lang, were distinguished in this field. Walton's first significant contribution to palaeobotany lay in the realm of techniques when in 1923 he described his 'transfer method' for the examination of both surfaces of delicate fossil plant material. Henceforth his 1920s published work included papers on the fossilised liverworts and mosses. In 1928 he described his 'peel method' for obtaining sections of petrified plant material. This method allowed serial sections to be obtained of structures as small as seeds, and dispensed with the more wasteful method of cutting and grinding. Further refinements of the method were described in 1930.

Whilst Walton came to Glasgow with an established reputation in palaeobotany, in some ways he arrived at an inopportune time. The Depression of the 1930s severely affected the finances of universities, and the effects were to be felt at departmental levels. Funds were generally limited, and even the smallest pieces of apparatus for personal research had to be purchased with money from external sources. Having weathered the Depression years, he was next faced with steering the Department through the War years of 1939-45 with all the attendant difficulties of falling student numbers and staff away on war service. Power cuts and poor heating had to be endured both during and after the War. The several curricula remained much as in Drummond's time, but with plant evolution, including palaeobotany, now a prominent additional topic. Walton planned the Honours course over a four-year period, and indeed made specific mention of this requirement in the University *Calendar.* The specialist courses for the fourth year were in the sphere of morphology, physiology and mycology, with some topics shared with agricultural botany for the Honours degree in Agriculture. As in Bower's time, botany remained a component of the M.A. degree. As with his predecessors, Walton was a member of three faculties, Arts, Science and Medicine. In common with Drummond, he was opposed to any moves to establish biology as a separate discipline, on the principle that botany would then lose its representation on the three faculties. In the post-war years additional staff members appointed included one with specific responsibility for agricultural botany, together with specialists on the algae, bryophytes and ecology. At the elementary level the departmental teaching requirements covered seven degree courses in Arts, Science, Agriculture, Medicine and Veterinary Medicine. Hence at second year level botany could be taken with a wide variety of subjects, some seemingly disparate. Class sizes were large at first and second year levels. Walton kept up the tradition of the Regius Professor lecturing to elementary classes. Although he lived nearby he was known on occasions to turn up for the 8am lecture minus a tie! Accommodation for plant physiology remained until 1937 restricted to a small room partitioned off from the workshop. A gift from Sir Daniel Stevenson, a Glasgow shipbuilder, enabled the building of an annexe, named the Stevenson Laboratory, for plant physiology. Special problems became acute with post-war staff increases, and in 1950 the spacious museum was in part converted into a number of laboratories and rooms for staff and advanced students, together with a seminar room which served as an extra lecture room, and a library.

Walton's palaeobotanical researches both before and during his time at Glasgow were wide ranging in terms of the geological ages covered, extending from the Lower Devonian up to the Cenozoic. He published on every major plant group with the exception of Angiosperms, with a particular emphasis on the Devonian and Lower Carboniferous periods. In later years he studied the Pteridosperms, fern-like seed-bearing plants, making notable contributions to knowledge of their seed structures. Whilst morphological research had continued in the Department a more experimental approach was developed. George Bond had been appointed Assistant by Drummond in 1927. Early on, under Walton, he was Carnegie Teaching Fellow before his appointment as Lecturer. His early researches were of an anatomical nature, mainly on the structure of the endodermis. In the early 1930s he commenced on an investigation of nitrogen fixation in angiosperms, so bringing the Department the first major moves towards experimental plant physiology. Bond remained in the Department until his retirement in 1976, being successively Senior Lecturer, Reader, Titular Professor and first occupant of the Hooker Chair of Botany in 1973. The researches of Bond and his associates, first on nitrogen fixation in leguminous plants and then from 1949 on root-nodulated non-leguminous plants, brought international distinction to him and to the Department, and to his election F.R.S., in 1972. Research on the fungi up to 1948 was mainly focused on fungal diseases of crop plants and trees utilised in forestry. After 1948 the theme was changed to an investigation of the production and roles of fungal antibiotics. Bryophytes were one of the main sources of taxonomic research, and ecological studies involved the marine algae and freshwater lochs. The bracken problem was also a focus of attention both during the War years and in the post-war period. Walton retired in 1962, having steered the Department through some difficult years, and seeing his staff double in number.

For his successor the University broke with tradition by going outside of academe. Whilst such a background was to some extent true of J. M. F. Drummond, he had been an academic for some ten years, interrupted by war service, before his move to the Directorship of the Scottish Plant Breeding Station in 1921. The 1963 choice for the Regius Professorship was Percy Wragge Brian F.R.S., a Cambridge graduate, a botanist and mycologist, who had charge of basic research at the Akers Laboratories of the ICI at Welwyn. Brian's main research interests were in plant pathology. In 1950 his research group at the Akers Laboratories followed up some earlier work in Japan in screening *Fusarium* spp. for a gibberellin-producing strain. One particular strain was found to be a good source and it was later shown that this fungal source gibberellin when applied exogenously overcame the dwarfism of a pea cultivar. This result led to a search for endogenous gibberellin-like substances in higher plants and Brian's group was one of two world-wide who demonstrated their occurrence. Hence Brian came to Glasgow with a well-proven record in both research and management. It did not escape the notice of his colleagues that he arrived in the Department on 1 April 1963. He arrived at a time when there was a fast-growing improvement in the financial climate of higher education, and with a recently appointed Principal, Sir Charles Wilson, who had a firm commitment to expansion of the University, and who had targeted the biological sciences as one of several areas for growth. With such a favourable environment Brian quickly realised that the Department was both under-funded and under-staffed with regard to current advances in the subject. In a relatively short time the staff number had doubled, with additional specialists in the field of ecology, taxonomy, plant pathology, genetics, physiology, biochemistry, cytology and fine structural studies. Combined with staff members already in post most of the major topics in botanical science were now covered for both research and teaching. There was also a marked increase in supporting staff, and a major replenishing of equipment and specialised apparatus. The growth in staff numbers placed a severe strain on accommodation, and a building in Lilybank Gardens was taken over as one outlet. For some years previously the possibility of establishing an outstation on the Garscube Estate had been under discussion and some forward planning was to hand, especially from the agricultural botany staff in the Department. Within two days of arriving in the Department, Brian had inspected the site favoured and had agreed that such a development was fully justified. He obtained financial support from the Nuffield Foundation and the Agricultural Research Council (ARC) for the building and equipment, including an electron microscope. A Unit for the Study of Obligate Parasitism was set up in the new laboratories, with Brian himself as its Director, and with staff funded by the ARC. In addition to the laboratories and staff rooms there were growth rooms and greenhouses. Another forward-looking move by Brian was in getting the Ordinance passed in 1965 for the foundation of the Hooker Chair of Botany, although another eight years were to pass before it was filled. Brian's reluctance to fill the Chair immediately was due to the lack of facilities then available which he considered would be an unfavourable environment for a new professor.

The growth in staff number and the enhanced range of specialisations led to a concomitant growth in postgraduate students and associates in addition to the ARC-funded staff at Garscube. Agricultural botany continued in strength both for teaching and research, with a teaching laboratory in the Chemistry

Department. Some storm clouds were gathering, however, over the future of the B.Sc. (Agriculture) at Glasgow due to a rationalisation programme under consideration, although the full effects of this were to fall on Brian's successor. Brian had been appointed not only to oversee the growth and well-being of the Department to keep pace with a rapidly evolving science but also to introduce the discipline of biology, the amalgamation which had proved such anathema to his predecessors in office. There remained some strong residual opposition to such a change, and in Brian's time the main planning stages were overcome, and again the full effects were a matter for his successor. Brian resigned in 1968 on his appointment to the Chair of Botany at Cambridge, taking with him the members of the ARC unit. His one regret from his time at Glasgow was that pressures of administration and external calls on his expertise allowed him no time for personal bench work.

John Harrison Burnett, an Oxford graduate, succeeded Brian in 1968. He joined the Department with a distinguished record of teaching and research gained at Oxford and Liverpool, and having held professorships at the universities of St Andrews and Newcastle-upon-Tyne. His principal research interests lay with the fungi, especially fungal genetics. Here he had made important contributions to knowledge of mating systems, mating-type factors, in- and out-breeding, distribution of breeding groups and inter-sterility of Basidiomycetes. In the year of his appointment he published *Fundamentals of Mycology*, a major monographic treatment. John Burnett arrived at a time of transition. The concept of a common biology course had been introduced at first-year (Ordinary) level, so replacing the separate elementary courses of individual subjects. A similar approach had been introduced for the first professional examination for medical students. For the Ordinary biology course it was envisaged that the identities of individual subjects should be less apparent although for lecturing purposes the representatives from individual departments covered those aspects akin to their own disciplines. In the second year Higher Ordinary stage the individual subjects were again the choice of the students. Another major change involved agricultural botany. The Bosanquet Report on the rationalisation of degree courses in Agriculture had recommended that Glasgow should be one of the three centres where the B.Sc. Agriculture should cease. In 1970 the B.Sc. in Agriculture became the B.Sc. in Agricultural Science within the B.Sc. Ordinance. This arrangement was repealed two years later and Agricultural Botany became an Honours Botany option within the B.Sc. Ordinance.

With the staff number little changed the overall concept of the Honours course remained much as before, covering the study of plants at cell, whole plant and community levels; plant physiology and biochemistry; plant pathology; plant ecology and geography; the morphological and ecological characteristics of Algae, Fungi, Bryophytes, Pteridophytes, Gymnosperms and Angiosperms; experimental taxonomy and genetics of flowering plants. For agricultural botany the topics included some shared with the above, and in addition the origins and history of crop plants; population genetics; crop improvement; utilisation and conservation of natural plant communities; crop ecology, and control of plant growth and development. Additional lecturing strength came with the appointments of specialists on the algae, and in the field of pollen analysis and vegetational history.

There were some inevitable teething problems with the Ordinary Biology course in its early years. There was seeming loss of identity with some disciplines, of which botany was one, with an attendant falling off in the numbers of students proceeding to the second year and beyond. Other problems lay in a lack of a centre for lectures and laboratory classes until the opening of the Boyd Orr Building. John Burnett played an important role in planning the Ordinary Biology laboratories in this new building.

Research at the Garscube Laboratory now included studies on the physiology and fine-structural aspects of hyphal growth and wall formation in fungi, and on experimental plant morphology both at the whole plant and fine-structural levels. Other features of mycological research in the Department included the role of fungal metabolites and aspects of plant pathology. In plant physiology the programme of research on nitrogen fixation continued, and was joined by investigations on translocation in plants. Work continued on the taxonomy and ecology of Bryophytes, and on the ecology of marine red algae. Research in agricultural botany included studies on seed germination processes and herbicide effects. John Burnett resigned in 1970 on his appointment as Sibthorpian Professor in the University of Oxford. In later years he was to return to Scotland as Principal of the University of Edinburgh, and to be knighted for his services to university education.

Malcolm Barrett Wilkins was appointed to the Regius Chair in 1970. A graduate of King's College in the University of London, his previous professorships had been in the universities of East Anglia and Nottingham, at the latter as Professor of Plant Physiology in the School of Agriculture at Sutton Bonington. He came to Glasgow with a distinguished record of research in plant physiology, especially on circadian rhythms in plants, and on the physiology and metabolism of plant growth substances. He also came with a great enthusiasm for teaching at all levels of student attainment, and within a few weeks of his arrival was lecturing to some 300 Ordinary Biology students. Whilst both the administration and logistics of this course were improved with the opening of the Boyd Orr Building, the availability of greenhouses in the roof area greatly facilitated both the supply of plant material and the prosecution of student experimental work. For some years various schemes for rehousing the biological sciences as a whole had been under discussion, including a 'greenfield' development of appreciable dimensions to house all relevant Departments with a common block for administrative offices and lecture rooms. P. W. Brian had been in favour of such a move, but this proposal did not appeal to Malcolm Wilkins and most of the departmental staff. The 1901 building occupied a pivotal position on the campus at Gilmorehill. Whilst some of Bower's professorial colleagues in 1901 had regarded his new building as altogether too spacious, this factor was to act very much in the favour of the Department in the 1970s. Better usage of available space was necessitated by an increase in staff number; an enlarged postgraduate school (which had more than doubled overnight with the arrival of the research group from Sutton Bonington); and the tide had turned with increased student numbers entering the Department at all levels. Hence came the need to restructure the interior. On the ground floor a laboratory was established for research on pollen analysis and vegetational history, and an enlarged laboratory on the second floor became available for research on the Algae, together with staff rooms. The courtyards between the Department and the Stevenson Laboratory were converted into a suite of controlled environment rooms and a workshop and a large preparation room. The spacious Herbarium of Bower's time became on the ground floor a large laboratory for physiological and biochemical research (named the Brian Laboratory) with staff rooms, and with a further suite of staff rooms and an area for the Herbarium on the second floor. The top floor teaching laboratories underwent modifications to give junior and senior Honours teaching laboratories with preparation and experimental rooms. The Higher Ordinary teaching laboratory was located in the Boyd Orr Building.

In the late 1970s the large lecture room became superfluous for University requirements, and this large space was converted into three floors. The top (second) floor was made into two lecture rooms, the larger named the Bower Room and the smaller the Walton Room. The middle or first floor was converted into two laboratories for postgraduates, the Hooker (after Sir William Hooker, Professor 1820-1841), and the Arnott (after G. A. W. Arnott, Professor 1848-1868). The ground floor became the Bond Laboratory for Higher Ordinary students, together with a common room for staff and students. The Garscube Laboratory continued as a centre now for mainly plant physiological research with all the advantages of gardens and greenhouses for the maintenance of experimental plant material.

By the mid-1970s the staff had increased to 18, with four to six research assistants and fellows regularly in post, and with a postgraduate school of some 20-30 students annually. The final Honours year was now regularly some 20-30 students, and as such probably represented one of the largest botany Honours schools in Britain. Agricultural botany remained a vigorous side, and its link with the West of Scotland Agricultural College retained via honorary lectureships for members of its staff. Teaching programmes in association with other Departments developed with two Higher Ordinary courses in Applied Biology and Marine Science, and at Honours level with Cell Biology. The wide ranging staff specialisations were also reflected in the forward looking courses for Ordinary degrees. Field courses remained an integral part of undergraduate teaching with visits to a number of field centres in Britain. A new departure was seen in introducing students to a distinctive overseas flora, with the result that an excursion to Tenerife is now a regular winter event in the junior Honours curriculum. Improved facilities for field work at both undergraduate and postgraduate levels came with the opening of the University's Field Station at Rowardennan on the banks of Loch Lomond, and the establishment in 1970 of the University Marine Biological Station at Millport on the Isle of Cumbrae in the Firth of Clyde, a venture jointly shared by the universities of Glasgow and London, and made possible by the removal of the staff of the Scottish Marine Biological Association from Millport to the newly opened Dunstaffnage Marine Science Laboratory at Oban.

The rapid growth of departmental research after 1970 and well into the 1980s lay in the experimental and ecological spheres. The principal thrust of the experimental research was in plant physiology and biochemistry. Plant hormones were a major field of research, including the roles of IAA, abscisic acid and ethylene in the growth regulation and differentiation of higher plants, especially in processes of branching and apical dominance, leaf senescence, dormancy, flowering and root production and nodulation. The isolation and characterisation of gibberellins, their biosynthesis and metabolism, completed an all round programme of research on plant hormones probably unequalled in its scope elsewhere in Britain. Other physiological topics included the study of circadian rhythms of CO_2 production and metabolism, and the sensory physiology of geotropic and phototropic responses; control mechanisms in seed germination and bud dormancy; and the mechanisms of translocation processes in phloem and xylem. Associated biochemical research included the role of short- and long-chain fatty acids in seed germination, and the function of endogenous plant steroids in plant physiology and development. The work on nitrogen fixation now involved some field-orientated aspects, notably with woody non-leguminous plants such as *Alnus* and the N-fixing actinomycete *Frankia*. Experimental plant morphology at both whole plant and fine-structural levels was a continuing field of research, including the modes of origin of leaves, buds and roots, and on processes of senescence. In the latter half of the 1970s some notable additional experimental fields became part of the broadening Departmental research school, namely, the primary reactions in bacterial photosynthesis, with particular regard to the light-harvesting processes and the full range of antenna-complex structure and function; the electron transport components involved in the transduction of light energy to chemical intermediates in the primary processes of photosynthesis in higher plants and algae; and the biosynthesis of higher plant cell wall polysaccharides, including the effects of plant hormones, and the control of cell wall extension.

Mycological research included a continuing programme on the nature and role of volatile metabolites, and an investigation of the effects of environmental factors on *Serpula lachrymans*. In the sphere of plant pathology, the potato blight fungus *Phytophthora infestans* was subject to intensive examination regarding the induction of susceptibility in potato tubers and the roles of preformed resistant factors. The functions of sterols in the reproduction of *Phytophthora* and the disease-producing genus *Pythium* and the genetics of viruses pathogenic on plants were other dimensions of plant pathology under investigation.

Ecological studies included the taxonomy and ecology of Bryophytes, the ecology of marine and freshwater algae, the ecology of freshwater macrophytes, and palaeoecology and vegetational history. The researches on algae involved the seasonal ecology and productivities of marine and freshwater phytoplankton; the intertidal and subtidal ecology of marine algae, especially the factors governing the distribution of fucoid algae; competition and colonisation by juvenile stages and the effects of grazing animals; pollution studies involved environmental impact assessments regarding the Hunterston iron-ore terminal;

other applied work included the corrosion effects of microbial marine algae on metals used in offshore structures (in collaboration with the Department of Mechanical Engineering). The palaeoecological studies on the statigraphic distributions of pollen and macroscopic fossils have concentrated on the late-glacial deposits in Perthshire and post-glacial peats and muds in west-central Scotland and have yielded much data on the vegetational history of Scotland. Archaeological sites have yielded information on plant exploitation for food and other purposes, including Iron Age material from the Orkney Islands, Roman material from the Antonine Wall, and medieval material from Perth. In the realm of aquatic ecology attention has been focused on the control and management of macroscopic plant communities in freshwater habitats - canals, rivers and lochs. Terrestrial weed control studies have been concerned with the responses to herbicides, and with weed/weed and weed/crop plant interactions.

Such wide ranging research programmes have been supported financially from a variety of sources, namely, the SERC, NERC, AFRC, MRC, the CEGB, the Forestry Commission, the Nature Conservancy Council, the Scottish Development Department, the British Council, British Steel, and the Marine Technology Directorate. Close working relationships have been established with institutes and university departments in Britain and the Commonwealth, and in the USA, South America, Scandinavia, Germany and the Middle East.

The financial insecurity which affected universities from the early 1980s with irregular patterns of shifting guidelines emanating overtly and covertly from official sources, was naturally felt at all departmental levels. For Botany, by the mid-1980s there had been some radical changes in staffing structure, with resignations due to promotions elsewhere and retirements of senior members of staff. Full-time staff membership was in consequence halved. A system of rotating departmental headships was introduced when Malcolm Wilkins demitted office on becoming Dean of the Faculty of Science, with Professor John Hillman, a plant physiologist, the first in post until 1986, when he left to become Director of the Scottish Crop Research Institute at Invergowrie. He was followed by Dr A. M. M. Berrie, Reader in Agricultural Botany, and his successor is the present Head, Professor Richard Cogdell, whose research interests are in molecular biology. Radical changes in teaching programmes have been made in the last five years. The second year now consists of three half courses - molecular plant science, environmental plant science and applied botany, so allowing student choice. The third year is a 'core' course, a common syllabus of direct relevance to present day developments. The fourth year is based on option modules of five weeks duration, choice of which can lie within the Department, or may extend laterally to other cognate modules in other departments within the biological sciences.

The organisation of Departmental research has also changed, with now four main areas, namely, molecular biology and plant biochemistry, plant physiology, ecology and palaeoecology, and plant pathology. In the field of molecular biology research on bacterial photosynthesis has led to the crystallisation of a purple bacterial antenna complex, the gene sequence of its formation has been worked out, and the regulation of expression is under examination. The light control of plant development is being studied at molecular level to determine receptor mechanisms, and the regulation of gene expression in relation to the signal transduction pathway. Research on plant rhabdoviruses has shown the presence of defective interfering particles in host plants which may influence the response of the plants to long-term chronic virus infections. The study of circadian rhythms in *Bryophyllum* has resulted in the isolation and characterisation of the enzymes concerned, and the factors regulating their activites are being examined. The photosynthetic processes and pathways in Cyanobacteria are under detailed investigation. The biochemistry of plant cell wall formation is a continuing study, with an interesting dimension in links with space research to examine the effects of microgravity on the process. Research on nitrogen fixation continues, some 60 years after first being studied in the Department. Current research centres on *Frankia*-nodulated non-leguminous plants, with a view to improving fixation properties by inoculating nursery grown *Alnus* with 'elite' strains of *Frankia*. Similar studies with *Casuarina* are exploring the possibility of improving its growth and so helping to limit desertification processes in Egypt. The possibility of extending *Frankia* root nodulation to non-nodulated woody species of economic importance is also under investigation. The biosynthesis and metabolism of gibberellins and auxin remain a prominent component of plant physiological research. In plant pathology the molecular approach with plant viruses has been described. The interactions between wild plants, such as the groundsel, and its fungal parasite *Erysiphe* are under examination and have shown that there are extremely variable responses, with a large number of specific resistance phenotypes, and the parasite shows equally variable specific virulence phenotypes. Research on freshwater ecology in relation to macroscopic plant communities and on wetland ecology is continuing and has been extended to overseas localities, including the River Nile. Collaboration with the N-fixation programme continues with a study of nitrogen inputs into wetland ecosystems. Similarly the palaeoecological and vegetational history research, and the archaeological studies, remain ongoing programmes. A link with the strong research tradition of the 1890-1925 period remains with current work on experimental morphology, with all its implications regarding horticultural practices. Members of the Department are actively involved in the University's Centre for Research in Environmental Science and Technology, a cross-disciplinary organisation set up to facilitate the formation of environmental consultancy and research projects. For some 30 years the Department was internationally recognised as a centre for research on marine algae. It is to be regretted that with staff changes of recent years this field of research is no longer practiced.

While present day trends in teaching and research show some departures from those of previous years, traditional aspects of botany have not been ignored. The 'New Botany' which Bayley Balfour and Bower brought to Glasgow was not intended to completely supplant the taxonomic approach but to complement it. Plant taxonomy - the science of classification - remained an essential background to teaching and research, as may be seen in the scale of the Herbarium in Bower's 1901 building, with its permanent demonstrations of plant evolutionary series always available for study purposes. Whilst reduced in scale, the present day Herbarium remains as a source of consultation and study for all. With all the recent curricular changes there is still a role for systematic and taxonomic studies. It is a well known maxim that a sound ecologist is also a practising taxonomist, and the main emphasis of field courses is in the identification of plants and communities. In 1987 the Department played a leading role in the Glasgow Botanical Expedition to Papua New Guinea, in collaboration with staff from the Glasgow Botanic Garden, an expedition sponsored by the Trades House of Glasgow, and under the patronage of the Royal Society of Edinburgh. Many species of plants (ferns, begonias and orchids) were brought back, with some new to cultivation. Stocks are held in the Botanic Garden, and are under investigation. Also in recent years a research team based in the Department has carried out a detailed survey of the flora of Glasgow in which traditional methods of identification have been linked with modern methods of data handling and retrieval. This will result in a new Flora of Glasgow, a unique contribution to urban ecology and conservation.

The beneficial mutual association between the Department and the Glasgow Botanic Garden can be traced back to 1818 when Robert Graham, the first occupant of the Regius Chair, helped plan the layout of the Sandyford Garden.

In 1820 Graham left Glasgow for the Regius Chair of Botany at Edinburgh, together with Keepership of the Royal Botanic Garden, and the title of King's Botanist in Scotland. Here he was soon involved in the removal of the Edinburgh Garden to its present day site at Inverleith. This 1820 link with the Royal Botanic Garden has recently been renewed with the appointment of the present Regius Keeper, Dr David Ingram, as an Honorary Professor in the Department and the appointment of Malcolm Wilkins as a Trustee of the Garden. For both Dr Ingram and the Department this is a gratifying connection. He was a member of the research staff at the Garscube Laboratory under P. W. Brian.

In 1893 Botany in the University had been updated by the introduction of the 'New Botany' to complement the traditional approach. In 1993, with the centenary of the Faculty of Science, we also look forward to the 300th anniversary of the commencement of botanical teaching in the University, fully aware of the dynamic nature of the subject, and having witnessed another 'New Botany' being introduced to once again update and complement, but not entirely supplant, an established system. This represents a striking example of new growth being grafted onto a sound stock, so ensuring healthy and progressive future development. An even more gratifying feature is the retention of the integrity of the Department, and of the title Botany, "the science which treats of plants" - within a school of biological sciences. Few such titled departments remain today in Britain, and none other in Scotland.

In conclusion it is perhaps fitting that we should look back with respect over some 289 years to the man who first taught botany in the University. John Marshall, 'Chirurgeon and Bottanist', left few tangible records, but his memory is perpetuated in the Marshall Room, where present-day students and staff meet for lectures and seminars.

Acknowledgments
I thank Michael Moss, Archivist to the University, for his assistance at all times; the help of my colleagues in the Department is gratefully acknowledged, especially that of Dr A. M. M. Berrie, formerly Reader in Agricultural Botany. I thank also Professors M. B. Wilkins and R. J. Cogdell for their helpful comments on the article whilst in manuscript form.

Arthur D. Boney is Emeritus Professor in the Department of Botany at the University of Glasgow.

Astronomy

Archibald E. Roy

The study of astronomy has a long history in the University of Glasgow. It was one of the subjects allotted to the Professor of Natural Philosophy and was included in the M.A. degree examination. The University had a telescope in 1693, two centuries before the creation of the Faculty of Science. In 1754 Alexander Macfarlane, merchant in Jamaica and brother of the Laird of Macfarlane, the antiquary, left to the University his astronomical equipment. This collection was shipped to Glasgow. Its original inventory, in French, is still in existence and of considerable interest - one of the listed pieces of equipment is 'One artificial eye'!

James Watt was employed by the University to put this collection in order, an observatory being built for it on the Dowhill in 1757 and named the Macfarlane observatory. The Dowhill, where, according to legend St Kentigern sat, stood near the south-eastern corner of the Old College grounds. "Well-equipped with astronomical instruments", as Tobias Smollett, himself a Glasgow student, put it in *Humphrey Clinker*, it was undoubtedly an excellent example of its kind. Towards the end of the 18th century, it possessed a reflecting telescope made by Sir William Herschel, the best telescope maker in Europe.

The Faculty of Arts, with the aid of the Duke of Argyll, founded a Chair of Practical Astronomy and Alexander Wilson, M.D. (1733-86) was appointed first Professor and Observer. His work on sunspots, leading to his discovery of what is still known as the Wilson effect, was awarded in 1772 the Gold Medal of the Royal Society of Sciences in Copenhagen. Among his successors were his son Patrick Wilson (1743-1811), the Professor of Natural Philosophy, William Meikleham (1771-1846), the Rev. James Couper, minister of Baldernock who on his death in 1836 was succeeded by John Pringle Nichol (1804-1859). Nichol was one of the most remarkable men who ever held a Chair in the University. He was not only a great observer but also a great teacher. His influence, as David Murray states in his *Memories of the Old College of Glasgow*, "was not limited to the students whom he taught, but extended to the whole students of the University as a body . . . His influence with the citizens of Glasgow was equally great. As a lecturer to a popular audience he was unrivalled."

The Dowhill observatory had a long record of useful work. Its observations were published in scientific journals. Its fate, however, was to fall a victim to the growing city's rising level of smoke pollution and to be surrounded by a number of tall buildings, restricting its view. Nevertheless, work continued as long as possible. Lord Kelvin was taught by Professor Nichol in the Macfarlane observatory and John Nichol recorded how "I used to mount up the rising ground on the College Green to the Old Observatory . . . and see my father with his fur cap on winter nights peering through the tube of the old telescope. I was alive to the wonders of the old tube. I remember seeing Orion through it for the first time, and Jupiter and the moon".

A second observatory had in fact been built on Garnethill between Scott Street and Thistle Street by 'The Glasgow Society for promoting Astronomical observation'. Under its foundation stone lay a plate dated 11 May 1810, stating that the building was "intended to promote the study of astronomy, and to record observations the most interesting, it is hoped will long remain . . . a magnificent monument of the scientific taste, and public spirit, of the City of Glasgow, and of the present time". For some years it was popular but ultimately a decline in interest set in. It was offered to the University of Glasgow but the offer was declined, partly because of the restrictive conditions attached, and partly because the University saw it even then suffering the fate of the Macfarlane observatory, surrounded and polluted in its seeing conditions by the fast-spreading city.

It was Professor Nichol's lecture courses in 1836 that inspired a new venture, set up by 'The friends of Astronomical Science in Glasgow' to erect a new observatory. A site in Dowanhill was purchased. The buyers were careful to exact a stringent obligation from the Dowanhill Trustees not to erect in the vicinity of the site any view-restrictive building or any smoke-emitting furnace. Those modern citizens of Glasgow familiar with the present Dowanhill-Hillhead-Hyndland area will find John Nichol's account of his first visit to the new observatory in 1840 of some interest.

"One afternoon in the autumn of 1840, I and my father and uncle William went to take . . . a very long walk into the country. We passed the outskirts of the town - then about Sauchiehall Street, left behind a few detached houses, which remain to recall the old days of St George's Road; crossed over the fields where Queen's Crescent was afterwards erected; and went on by an ill-made narrow road that skirted the great black quarry; and reached a farm-house on Horselet Hill, and plunged through ploughed fields to the top, where a crowd of masons were planting the foundations of the future Dome."

Professor Nichol took up residence in the observatory in April 1841. Unfortunately the institution, like its Garnethill predecessor, ran into serious financial difficulties. The University of Glasgow agreed to take over the observatory, pay its debts, and the Old Dowhill equipment was as a consequence transferred to Dowanhill. It may be noted that in the University of Glasgow's Bute Hall, a beautiful window, presented by Professor Jack, commemorates Professor John Pringle Nichol, his son Professor John Nichol and his daughter, Mrs Jack.

Robert Grant succeeded Professor Nichol, being appointed Professor of Practical Astronomy in 1859. In the winter of that year he carried on the class of natural philosophy when Lord Kelvin met with an unfortunate accident. Grant, born at Grantown-on-Spey in 1814, was not only a noted scholar on the history of astronomy but also a practical solar researcher. In 1852 he was appointed editor of the *Monthly Notices of the Royal Astronomical Society,* a task to be relinquished on his appointment to the University of Glasgow.

Grant was a worthy successor to Nichol. To the instruments he found at Dowanhill he added, in 1863, the Ochtertyre equatorial telescope and a small three-inch aperture transit telescope. Over the years of his professorship he extended Nichol's stellar work, observed minor planets and published in 1833 *A Catalogue of 6415 stars for the epoch 1870, deduced from observations made at the Glasgow University Observatory during the years 1860-1881. A Second Glasgow Catalogue of 2156 stars* was published just a few weeks after his death on 24 October, 1892. Although an invalid by that time he examined every proof-sheet in the catalogue, looking over the last sheet on the morning of his death.

Other notable work carried out at the Dowanhill observatory in the second half of the 19th century included the setting up of a system for controlling a number of public clocks in the City and Port of Glasgow by direct current from the observatory so that they might show correct Greenwich Mean Time. The system, once instituted, continued as a time-keeping service far into the present century. The observatory, in 1868, was chosen as a principal meteorological station by the Meteorological Office in London. Again, this service, once begun, was continued until the observatory was demolished in 1936.

Grant, in addition to his lecturing schedules, carried out all this work with the aid of one Assistant, Mr James Connell. In recognition of his work, he was awarded two honorary degrees; he was elected a Fellow of the Royal Society in 1865 and he received the Gold Medal of the Royal Astronomical Society in 1856 for his *History of Physical Astronomy.*

It is tempting to suggest that the history of the University of Glasgow's Department of Astronomy in the century from the creation of the Faculty of Science in 1893 to the present can, like Caesar's Gaul, be divided into three parts, namely the Becker, Smart and Sweet incumbencies in the Regius Chair of Astronomy, for their span of years occupies all but six years of the Department's history in that century.

Ludwig Becker was appointed to the Regius Chair in January 1893, at the age of 33, a position he was to hold for 42 years. He was a highly experienced practical astronomer and a pioneer in spectroscopy. Additionally, he was a first class mathematician and quickly proved himself to be an inspiring lecturer. Born in Wesel on 8 January 1860, he studied mathematics and physics at Berlin University. From 1879 to 1881 he studied at Bonn, specialising in astronomy. In 1885 he accepted a post at Lord Crawford's observatory at Dunecht, Aberdeenshire. There he extended greatly his experience as an observer, making spectroscopic observations with the 15-inch refractor. He also completed a series of observations of some 200 nebulae. For his work in observing the solar spectrum near sunrise and sunset through three summers from a hill - the Barmerkin of Echt - he was awarded the Macdougall-Brisbane prize of the Royal Society of Edinburgh.

In 1889, Dr Copeland, the Director of Lord Crawford's observatory, was appointed Astronomer Royal for Scotland and Becker became Assistant at the Royal Observatory. Much of the preparations for moving the instruments from Dunecht to Edinburgh fell to Becker and in particular he was responsible for adjusting the piers to support the transit circle. While the work was still in progress he was appointed Regius Professor of Astronomy at Glasgow.

His first task was the renovation of the instruments. For this work he was fortunate in receiving a grant of £1,000 from the Bellahouston Trustees. The 6-inch Ertel transit circle was reconditioned and the 20-inch Breadalbane reflector was overhauled and provided with a large spectrograph, a really powerful instrument for that time. He started observing as soon as the instruments were ready. With the transit circle his programme was to observe stars close to the pole and in particular two faint stars so close to the pole that they were always within the field of view of the telescope. By observation of the same stars at intervals of several hours he was able to determine the position of the pole relative to the instrument and then accurate positions of the stars. His intention was to determine accurately the constants of nutation and aberration. Although hampered by the unsuitable climate and the neighbourhood of a large industrial city, he carried on the observations single-handed for ten years. The results were published in the *Memoirs of the Royal Astronomical Society.* With the spectroscopic observations, weather was also a most disturbing factor but very long exposure spectrograms were made of novae, comets and nebulae. Spectra of the major planets were also photographed. Some results were published in the *Monthly Notices of the Royal Astronomical Society.*

Becker was a most approachable professor, not only to his own students but to others within and without the University. As a student, Alexander Thom, who in later life became Professor of Engineering Science at Oxford University and renowned internationally as a pioneer of the discipline of archaeo-astronomy, came to Becker to seek his advice in computing orbits. Ever afterwards he recalled Becker's friendliness and brilliance in grasping and solving the particular problem Thom was faced with. Soon after term began, Becker would give his students a key to the observatory and tell them that the observatory was open from "noon till noon" - the limit of the astronomer's day before 1925. For most of his career he probably had larger classes than any other professor of astronomy in the UK. His lectures were not based on popular or merely descriptive astronomy but contained a sound theoretical development of spherical and dynamic astronomy. Students were taught how to compute the ephemeris of a planet from the orbital elements and made to do so; in practical astronomy they had to learn to observe with a variety of instruments and determine clock error, latitude and azimuth.

The outbreak of the First World War began a period of considerable sadness for Becker. The yearly Glasgow observatory reports in the *Monthly Notices of the Royal Astronomical Society* are almost unchanging in their content from year to year during the War, being short and devoid of the enthusiastic descriptions of instrument-refurbishing and research achievements of pre-war years. For the duration, they blandly state that "The Director has been on leave of absence. The usual routine work connected with the time service and meterological observations has been carried on by the assistant". In fact, although Becker had long been a naturalised British subject, most of his relatives were in Germany and his absence from Glasgow was enforced upon him as a result of local agitation, a mindless and vicious xenophobia still existing in our present day.

Becker was entering his sixties when he returned. Too old to undertake regular night observing as well as carrying out his day duties, he concentrated on teaching, innovations dealing with equipment such as improved sextants for use at sea or in the air, and theoretical research. It was not merely a matter of age and his numbing war-time experience that produced this change in direction. As an observational observatory, Dowanhill was suffering the fate of its predecessors on the Dowhill and on Garnethill.

In his memoir *The Climatology of Glasgow,* published in 1925 by His Majesty's Stationery Office for the Meteorological Office, Becker laments how "The Observatory . . . has been encircled more and more in the course of time by buildings which, from the east round by south to west, now extend to a distance of from two to ten miles. In the other directions the houses in the immediate neighbourhood are detached villas." He might have added that increasing air pollution from factories had rendered serious observing totally impractical. From then until he retired in 1935 he carried on successfully with his teaching and theoretical research but the observatory, now used almost entirely for practical instruction, grew ever more dilapidated. Whoever succeeded Becker, he was going to have to do some hard thinking concerning astronomical policy in the University of Glasgow. The University itself, as will be seen, had its own radical thoughts on the matter.

Ludwig Becker's successor in the Regius Chair was William Marshall Smart. He came to the Chair of Astronomy in 1937 with a distinguished academic and research reputation. He was not yet fifty years of age. Born in Doune, Perthshire on 9 March 1889, he had been educated at McLaren High School, Callendar before proceeding to the University of Glasgow where he read mathematics, natural philosophy and chemistry. He received the Cunningham Medal for Mathematics and the Breadalbane and Ferguson Scholarships. Among his classes was astronomy, his teacher being Ludwig Becker. Becker's skilled exposition of the ways mathematics served as the handmaiden of a number of astronomical disciplines must have aroused Smart's interest for on graduation from the University of Glasgow in 1911 with First Class Honours, he became a scholar of Trinity College, Cambridge and holder of the Sheepshanks Exhibition in Astronomy. There he read pure and applied mathematics (including astronomy) obtaining a first in Part I of the Mathematics Tripos and becoming a Wrangler in Part II. He graduated in 1914, was awarded the Tyson Medal and in 1916 received the Rayleigh Prize.

The First World War intervened in his career. In 1915 he entered the Royal Naval College, Greenwich and a year later entered active service until the end of the war. As Instructor-Lieutenant, he served on *HMS Emperor of India,* the flagship of a division of battleships of the Grand Fleet. To the end of his long life he was able to reminisce with affection about his naval experiences and sometimes, during his lectures to University of Glasgow undergraduates, allusions to navigation, naval strategy and sea battles would lighten for his students stretches of spherical trigonometry or discourses on the combination of observations.

An early interest in celestial mechanics was temporarily sidetracked by the First World War. A paper on the Trojan asteroids (published in 1917) was followed in 1921 by a paper on the anomalous rate of revolution of Mercury's perihelion and one in 1925 on the evolution of binary star systems as a result of mass-loss by radiation. In fact, in the 1920s, it seemed that the great days of celestial mechanics were over; the Hill-Brown lunar theory seemed the culmination of work in that field and theories of the planetary movements left little to be desired. Nevertheless, one of Smart's most successful books was his *Celestial Mechanics*, published in 1953 and still being used in the late 1980s as a text book and a work of reference. As in all his work it exhibited his rigorous mathematical training.

The First World War gave him a life-long interest in navigation. A paper in the *Monthly Notices of the Royal Astronomical Society* in 1919 on the simplification of tables for nautical navigation was followed in 1922 by the *Admiralty Manual of Navigation* and in 1924 by *Position Line Tables*, both co-authored by Cdr. F. N. Sharme, R.N. In the next 22 years he wrote four books on navigation and during the Second World War he taught navigation to RAF cadets taking crash courses at the University of Glasgow. No doubt he read with much interest the paper *On a New Aircraft-sextant for Use with Visible Horizon*, published in 1936 in the *Monthly Notices* by the Regius Professor of Astronomy in the University of Glasgow, his predecessor.

Perhaps his most famous book was *Spherical Astronomy*, first published in 1931. There are probably few university or observatory libraries in the world that do not have a copy of one of the many editions of this jointly successful work on their shelves. Generations of astronomers, navigators and university students have used it to master the intricacies of celestial coordinate systems and time and grasp the many subtle ways in which precession, nutation, aberration, parallax and refraction can tinker with the directions celestial objects seem to lie in. Only relatively recently has it had to be at first up-dated and then replaced by Robin M. Green's text of the same name to take into account the much greater accuracy to which these matters are observed in our modern space age.

After the First World War, Smart became John Couch Adams Astronomer and Chief Assistant at the University Observatory, Cambridge, which was then under the direction of A. S. Eddington. The post not only involved theoretical research but also carried an obligation to do practical work using the Sheepshanks 12-inch Coudé refracting telescope. A series of papers on photographic photometry using sodium/argon and potassium/argon photoelectric cells was published between 1921 and 1937. Much of this pioneer practical research work was carried out on variable stars; the papers exhibit Smart's cautious, painstaking attitude in evaluating and taking into account error sources. His 1958 book *Combination of Observations* shows how experienced he was in practice as well as theory.

Further research work involved a long programme designed to study proper motions. With H. E. Green he used a set of parallax plates taken with the Sheepshanks telescope in 1903-04 as a set of first-epoch plates. The second-epoch plates he took with Green were exposed through the glass so that the first and second epoch plates could be compared 'back-to-back'. A careful discussion of the reliability of this method appeared in a 1921 paper, with proper motions determined for the Pleiades. It was followed in the *Monthly Notices* by ten papers, his last research paper in this topic appearing in 1941.

Perhaps the best example of his patient, painstaking and honest approach to laborious, important, yet non-spectacular research was his series of over 30 papers on stellar kinematics, published between 1923 and 1940, occasionally co-authored with H. E. Green (Cambridge), T. R. Tannahill (Glasgow) and S. Chandrasekhar (India and the United States). When the series began, it was only a decade since the existence of the galaxy as a relatively small entity in a huge universe had begun to be established. The dynamics of stars pursuing their colossal orbits about the galactic centre was only beginning to be understood. Various models of systematic trends in stellar motions such as the two-stream theory and the ellipsoidal theory had been proposed. Smart's work over two decades collected and analysed tens of thousands of proper motions. Apart from the series of papers, he wrote in 1938 a book called *Stellar Dynamics* which became a much-used standard text. 30 years later it was up-dated and published under the title *Stellar Kinematics.*

The person who came to Glasgow to occupy the Regius Chair in Astronomy in 1937 was therefore by any standards a man of high attainments not only in practical and theoretical research in astronomy but also in navigation and mathematics. He was also a writer of great distinction, noted for his clear exposition of recondite matters. He had a well-deserved international reputation.

The situation he found in Glasgow was a challenging one. It was also a source of great disappointment to him. The old Dowanhill observatory was now a century old. Built originally outside the city boundaries, it was now surrounded by the suburbs of Hillhead and Dowanhill with consequent major light pollution, apart from the gross deterioration in seeing conditions due to the increasingly filthy atmosphere of the industrial city. In the latter years of Professor Becker's directorship, the scientific work of the observatory in meteorology and astronomy had come almost to a standstill. The buildings themselves were in a poor state of repair and there is no doubt that astronomy in Glasgow was at its nadir.

The situation was such that from certain quarters in the University considerable pressure was applied to have the Professorship of Astronomy downgraded to a Lectureship. It was argued that Astronomy was not in a fit state to be an autonomous university discipline.

Smart met the challenge and ultimately won though, as Wellington remarked on a previous occasion, it was a damn close-run thing. The Chair was saved; the old observatory was abandoned. The University sold the site and today, all that remains of the third Glasgow observatory is the street name 'Observatory Road' in Hillhead and some instruments, including two telescopes and two clocks, now at the Garscube observatory. Indeed one of the telescopes, the transit circle, is still in use for student training purposes.

If politics is the art of the pragmatic, Smart's solution was truly political. Believing that serious observational work was now impossible in the vicinity of an industrial city such as Glasgow, he accepted that the future of the Department of Astronomy lay in teaching and theoretical research. The University was persuaded to use a fraction of the proceeds realised from the sale of the Dowanhill site to build a smaller observatory in University Gardens at the main University site in Gilmorehill. Designed mainly for instructional purposes, the substantial building (strong enough to take a second storey if a future need arose) included a large room to house the astronomical library from the old observatory and to provide accommodation for lectures. Generously-sized rooms were also provided for the Director and his Assistant, Dr T. R. Tannahill. There was also an instrument room. On the high ground behind, a dome housed the 7-inch refractor and a smaller building held the transit circle. The Riefler clock, keeping sidereal time, stood in its glass case in the hallway of the new observatory house, sending its pulses to a chronograph in the instrument room. A set of concrete and brick flat-topped pillars about one metre high stood in the grounds at the back. Troughs containing mercury could be placed on these pillars to provide artificial horizons when sextant observations of the sun were made by students measuring the observatory's latitude and longitude. It is interesting that no room was provided in the new building for a secretary or typist. Professor Smart invariable wrote letters by hand. As much as anything else it was a matter of courtesy to do so.

The new observatory was officially opened on 17 April 1939 by Sir Arthur Eddington, O.M., in the presence of a representative assembly presided over by the Chancellor of the University, Sir Daniel Stevenson, Bart., accompanied by the Principal, Sir Hector Hetherington and Professor John Walton, Dean of the Faculty of Science. During the proceedings Sir Arthur gave an address on 'The Expansion of the Universe'.

Teaching and research work were already gathering pace. Smart was not only beginning to make his mark on the undergraduate teaching but had also formed a small graduate school including Messrs. Akbar Ali, Archibald Brown and P. C. Chaudhuri carrying out research on various problems of stellar kinematics. Six papers published in that year by Smart and one by R. A. Robb and T. R. Tannahill indicate a healthy situation.

The outbreak of the Second World War necessarily curtailed the Department of Astronomy's activities and effectively ended Smart's own research career. Although, after the War, Smart received at Glasgow European astronomers such as Randic (Yugoslavia), Strassl (West Germany) and Eichhorn (West Germany) for research visits, his own publications, apart from textbooks and popular works, ceased to appear. Until he retired at the end of September 1959, almost all research in the Department was conducted by other members of staff and graduate students, an increasingly larger fraction of that research being on topics outside Smart's customary fields. Thus while Dr T. R. Tannahill and Mr D. G. Ewart continued to work on proper motions, Mr P. A. Sweet, who joined the staff in October 1947, was engaged in the investigation of sunspot magnetic fields and, aided by Mr A. E. Roy, was by 1952 researching a number of features associated with rotating stars. In September 1953, Sweet left to take up the post of Assistant Director of the University of London Observatory, being succeeded by Mr Michael W. Ovenden.

Two features of this immediate post-war period are of interest. The numbers of undergraduate students taking astronomy showed a marked increase and a growing number of members of the University and of various societies visited the observatory on Thursday evenings during the two winter terms to view the heavens, weather permitting. The Ordinary (first year) class in Astronomy approached and surpassed the one hundred mark in numbers and the number of undergraduates professing Honours in Astronomy or Astronomy with Natural Philosophy or Astronomy with Mathematics also grew to unprecedented numbers.

It was during Smart's last years of his tenure of the Regius Chair that he wrote many of his textbooks (*Celestial Mechanics, Foundations of Analytical Geometry, Combination of Observations*) and more popular works, among them *John Couch Adams and the Discovery of Neptune, The Origin of the Earth, Some Famous Stars,* and *The Riddle of the Universe.*

He was also much sought after as a lecturer so that in his popular books and lectures he followed closely in the steps of his illustrious predecessor, the great populariser of astronomy, Professor John Pringle Nichol. Nichol's books such as *The Architecture of the Heavens* and *Cyclopaedia of the Physical Sciences* could be seen in the libraries of most educated Glaswegians in the middle of the 19th century while his courses of popular lectures created such enthusiasm that the initial steps were taken to build the old observatory at Dowanhill.

A measure of Smart's skill as a lecturer is evident in the list of his official Lectureships. Apart from his undergraduate lectures he was Thomson Lecturer, Aberdeen; Elder Lecturer, Glasgow; Fison Memorial Lecturer, Guy's Hospital, London, and Halley Lecturer, Oxford. Many courses of lectures were given to the general public, not only in Glasgow; during the Second World War he had lectured to the Armed Forces. His immense efforts in the popularisation of astronomy were recognised by the award to him of the Lorimer Medal of the Edinburgh Astronomical Society.

Further measures of his distinction was his Fellowship of the Royal Astronomical Society (he was Secretary from 1931 to 1937, Vice-President from 1937 to 1938 and President from 1949 to 1951) and his Fellowship of the Royal Society of Edinburgh (Council member from 1949 to 1952 and from 1957 to 1959 and Vice-President from 1952-1955).

When Smart retired at the end of September 1959 he left a department which, though small in numbers of staff and students compared to other science departments such as Natural Philosophy and Chemistry, was healthy in teaching and research, and was expanding steadily. It was thus poised to take full advantage of two strong unrelated developments, the flood of new discoveries in astronomy, partially due to the advent of space research, and the expansion of higher education.

His successor in the Regius Chair of Astronomy was Peter Alan Sweet who had already served as a member of staff from October 1947 to September 1953. A Cambridge graduate and a brilliant mathematician, he had taken his Ph.D. under the supervision of Fred Hoyle and had had close contact with Hoyle's collaborators Hermann Bondi and Thomas Gold. In fact Sweet's pioneering work on the theory of magnetic reconnection in solar flare plasmas subsequently established the direction of some of the main research developments in plasma astrophysics. He had also demonstrated his ability as a Lecturer at Glasgow and at Mill Hill University of London Observatory when he was Assistant Director and so his appointment to the Regius Chair of Astronomy in Glasgow was welcome.

The Department Smart's successor came to in 1959 had 95 first-year, 14 second-year and four Honours students. Apart from the Director, who continued his research in magnetohydrodynamics, Dr Ovenden, in collaboration with Dr A. Wasserstein of the Greek Department, was researching the origin of the stellar constellations. Together with Dr A. E. Roy, who had replaced Dr T. R. Tannahill on his resignation as Lecturer, they were continuing their work on the dynamics and stability of the solar system, a work which became a major research theme over the next 30 years. Dr Roy was also conducting research (in mathematical biophysics) on the storage of information in neural networks. He and Mr J. S. Griffith, who had joined the Department as an Assistant Lecturer, were running an active Moonwatch team, made up of volunteers who on every clear night operated a 'fence' of 12 telescopes, with auxiliary recording equipment, set up to track artificial Earth satellites; measurements of the sightings of these satellites were sent regularly to the Smithsonian Astrophysical Observatory where they were used to update the predictions of the satellites' positions. In addition, Mr R. M. Green had joined the Department as a research student and Mr E. Edgar was appointed as a mechanic.

The '60s saw major developments in the teaching, research and equipment of the Astronomy Department. In the academic year 1961 to 1962, the staff complement consisted of Professor Sweet, Dr M. W. Ovenden, Dr A. E. Roy, Mr R. M. Green and Mr R. B. Hunter. Miss L. C. Muir joined the staff in October 1961 as departmental Secretary. There were 115 first year students, 12 second year, eight Honours students and two research students, placing the Department among the biggest institutions teaching astronomy in the UK.

The building in University Gardens had not been built for such staff numbers let alone student numbers. For lectures and tutorials the first-year class could still be accommodated in the old astronomy classroom in the Gilbert Scott Building and the other smaller classes in the observatory. For staff accommodation, however, changes had to be made. The original over-generous provision in 1936 of very large rooms in the observatory for the Director and his Assistant, and for lectures, now paid off. An ingenious re-planning of the observatory interior by Professor Sweet provided a director's room, two small staff rooms, a lecture room, and two small alcoves for research students to work in, the original Assistant's room now being shared by two members of staff. The old instrument room was divided into a workshop and a store-room. While it was a satisfactory solution at the time to the accommodation problem, it was recognised it could not be a permanent solution. Any further expansion of staff and the provision of a modernised practical undergraduate course involving fresh and more versatile equipment of observations and laboratory work required a fresh start. The University Gardens building had served its purpose and although it had been built originally strong enough to take a second storey, that was not the answer. And so planning began for a new observatory to be sited on the University of Glasgow's estate at Garscube, Glasgow, near Maryhill. It would become Glasgow's fifth observatory following the Dowhill, Garnethill, Dowanhill and University Gardens establishments. The University Gardens site, after demolition of the observatory building, dome and transit circle would have the Queen Margaret Students Union built upon it.

At this period, in the late '50s and early '60s, the influence of the changing scene in astronomy and the advent of space research was apparent in the Department's research. Visits were made by Dr Ovenden to the Dominion Astrophysical Observatory, Victoria, Canada, to take spectra of certain spectroscopic binary stars with the 72-inch telescope. He also spent seven weeks at the Kitt Peak National Observatory, Tucson, Arizona, making photoelectric observations of the star 57 Cygni with a 16-inch reflector. The observations were subsequently reduced using the University of Glasgow's KDF9 computer.

Dr Roy visited the Smithsonian Astrophysical Observatory and Goddard Space Flight Center, Washington, for research discussions. Under a grant from the Department of Scientific and Industrial Research he began a theory of the orbits of close lunar satellites, taking into account the evolution of such orbits under the gravitational influence of the sun and the Earth and the departure of the moon's shape from a perfect sphere. Dr Green took spectrograms of the sun at the University Observatory, Oxford, to obtain a test of a theory of the limb effect by comparing the centre-limb red-shifts of Swan bands of molecular carbon relative to atomic lines.

The Director, in collaboration with Dr Green, carried out theoretical work on the production of sheet currents in hydromagnetic systems. He also worked on instabilities in stellar systems and the stability of magnetohydrostatic systems in relation to the solar flare problem. An account of some of his work on the origin of solar flares was given at the AAS-NASA Symposium on the Physics of Solar Flares, Washington in October 1963. He was helped in this work by Mr D. MacGregor, who had come to the Department as a research student from the Department of Mathematics at the Royal College of Science and Technology, Glasgow. During this period, Dr Roy completed a second section of his work on the storage of information in neural networks and continued the work in collaboration with Dr Ovenden.

Mr R. B. Hunter and Dr Roy investigated the stability of Jupiter's outer satellites and by extensive use of the University's computer, established how asteroids could be captured by Jupiter as temporary satellites then released to re-enter the asteroid belt or be slung further out in the solar system. Two research students, Mr R. C. Smith and Mr W. McD. Napier were engaged in research into solar magnetic fields and binary stars.

By 1964, the Department had vacated the University Gardens site and had been placed in temporary accommodation in Ashton Road. It occupied one of the houses which were subsequently demolished when the University Avenue-Ashton Road area was modified. The only trace of the Department's occupancy is a tree still standing in what remains of Ashton Road, known to those who remember her as 'Miss Muir's tree' because she is said to have lobbied for its retention when the house was demolished. It still flowers each summer.

By 1970 a new and fruitful chapter had begun in the Department's history. After a rather longer time in the Ashton Road temporary accommodation than expected, the Department found itself in possession of a brand-new observatory in the Garscube estate. Part of the staff complement was also housed on the top floor of the new Department of Mathematics building stretching along one side of University Gardens, the Astronomy quarters actually lying across the Gardens from the site of Professor Smart's observatory.

The new observatory, on the outskirts of the city, near one of the University's Halls of Residence, was formally opened in March 1969 by the Astronomer Royal for Scotland. A combination of its choice of site, the Clean Air Acts and the decline of heavy, smoky industry in Glasgow had provided a marked improvement in 'seeing' conditions for Glasgow astronomy. The heavy, dirty, killer fogs still remembered by Glasgow's older residents were a thing of the past. The present author still recalls vividly how frequently because of fog he had to send his students home in the evenings an hour early from tutorials in the Gilbert Scott Building and then fumble his own way through the damp throat-stinging, confusingly thick fog to his own home, the nearby street-lamps dimly glowing in the murk like stars barely illuminating the dense enveloping nebula they were embedded in.

A great deal of careful planning had been put by the Department staff into the new observatory, its equipment and its teaching laboratories. It was equipped to provide student practical information and to support research in the field of instrument design and preliminary testing at a small telescope. The capital equipment comprised a 20-inch reflecting telescope by Grubb Parsons, a Joyce-Loebl microdensitometer and an Askania irisdiaphragm photometer. The building was designed to house a coelostat with a high-dispersion tunnel spectrometer. These facilities were supported by a fully-equipped workshop, library, lecture room and a Goto-Eros planetarium, teaching and research laboratories and an on-line connection to the KDF9 computer in the University.

A drive and guidance system was created for the 20-inch telescope. The basic drive was provided by a heavy-duty stepping motor. For sidereal rate, the motor required to be driven at 200 pulses per second. This frequency was obtained from a quartz crystal oscillator and electronic dividing circuits. By using a hand control, guidance in right ascension could be applied. Push buttons operated relays which controlled the divide circuitry, and the rate at which the stepping motor was driven was altered accordingly. Slow motion in declination was provided at two rates and the position of the secondary mirror could also be controlled by an oscillator whose frequency could be adjusted by the observer at the telescope. Some simple instrumentation for the telescope was also constructed. The equipment available comprised an attachment for a 35mm camera, a single-channel photoelectric photometer and an experimental photometer using a type of photodiode which had a voltage to frequency converter in its encapsulated shell.

The new teaching laboratories provided a long-sought opportunity to extend the range of undergraduate teaching in observational astronomy the Department had previously provided. Dr David Clarke, who had joined the Department as Lecturer in January 1967 from the University of Hull, was mainly responsible for much of the new development. In subsequent years, as Deputy

Director of the observatory, he continuously up-dated these teaching provisions, skilfully utilising innovations in equipment to let students see all aspects of any particular astronomical measurement. The general aim was for the student to collect her/his own data, reduce it and then interpret it. For this purpose, in some of the teaching laboratories, a series of star simulations was designed and equipment provided so that the experiments could be performed during the day and independent of the weather. There were artificial stars which lent themselves to visual observation and the determination of a double star orbit. There was a star field which could be photographed and the resulting plate compared directly with the real positions of stars in a catalogue. There was also a star whose brightness could be controlled, allowing photoelectric observations to be made and a light curve obtained which was then suitable for analysis. Other equipment in the laboratories was fairly standard and provided experiments for teaching basic optics or for analysing astronomical photographic plates.

The scope and range of the laboratory teaching when the new observatory was opened can be summarised as follows:

First year: The students had two demonstrations in the planetarium and performed quantitative exercises on the Earth's precession and on the motions of the planets. Other experiments and exercises included: the determination of the resolving power of a telescope: the determination of stellar magnitudes by diameter measurements on photographic plates: the investigation of the distribution of Wolf-Rayet stars and the consideration of galactic coordinates: the determination of the parallax of an open cluster by finding the convergent point: the investigation of the usefulness of the Cepheid and RR Lyrae variable stars for distance determinations: the observation of lunar positions during the course of the winter to determine orbital elements: and sextant observations of the sun in summer.

Second year: Besides the standard optical experiments on prisms and diffraction gratings, usually found in physics laboratories, the students performed practical work and exercises on: blackbody radiation: double star measurements, followed by Zwiers' graphical method: star field photography, followed by positional determinations: a simulated Michelson stellar interferometer: stellar radial velocity measurements: line profile measurements by microphotometry by a spectrum recorded on a photographic plate: and transit telescope observations when weather permitted.

Third and fourth years: The students performed the practical work on a project basis, some of which could last a number of weeks. Some of the projects consisted of: the determination of satellite orbits by photographic observation - the production of a Hertzsprung-Russell diagram from the analysis of plates taken in two colours: the measurement of equivalent widths of solar spectral lines by recording the spectrum photographically and performing microphotometry; the determination of the height of lunar mountains by photography and subsequent micro-densitometry: the determination of the selectivity of a telescope-photometer system: the observation of a light curve (artificial star) and the subsequent analysis of an eclipsing binary system.

Academics are singularly privileged in their continual encounter with each young generation as it arrives at university, demonstrating that the falsest of all myths is that 'the younger generation is not what it used to be'. The present writer, incidentally, once read that that statement was translated from a cuneiform clay tablet dug out of the ruins of King Ashurbanipal's library at ancient Nineveh! Over the past 40 years, an increasing number of students, undergraduate or postgraduate, entered the Department of Astronomy, graduated B.Sc. or Ph.D., some of them spending a further number of years doing research before seeking new posts elsewhere, many of which, it is satisfactory to note, lay in astronomy or space research scattered round the world.

In the '70s and early '80s, members of the Department markedly increased their number of research collaborations with colleagues outside or in the UK. India, Brazil, France, Canada, Greece, Japan, Spain, Italy, the United States, and the Netherlands were among the countries involved in such collaborations. A number of foreign senior scholars and graduate students from some of these countries arrived for extended visits to conduct research at the Department, their presence and input greatly enriching the lives of its staff and students.

While there was some decline in first-year student numbers to new competing subjects during the '70s and '80s, it was a period that saw the size of the Honours class treble and graduate student numbers quintuple, changes associated with new appointments, diversification of research and departmental restructuring. These changes are too numerous to detail in such a short history of the Department and can only be chronicled briefly below. Alongside these, the Department gave hundreds of extramural and schools lectures, many under the auspices of the Royal Society of Edinburgh Cormack Committee, and organised numerous public visits to the Garscube observatory including Halley Comet evenings during the 1986 apparition. In the late '80s, the 20-inch research telescope was moved to a new site on the University's Cochno Estate to escape the increasing city light pollution, the main teaching facilities remaining at Garscube. Other major teaching activities were service lecture courses in the departments of the History of Science, Geography/Topographic Science, and Aero Engineering, and, in W. M. Smart's footsteps, publication of five well received textbooks by A. E. Roy, by A. E. Roy and D. Clarke, by R. M. Green, and by I. J. D. Craig and J. C. Brown.

To appreciate the success of Glasgow astronomy in recent times, it is useful first to comment on the trying times in which this progress was set, with chill winds of inflation, of recession, and of government policy unsympathetic to education and research. Research councils moved from funding all 'alpha' and some 'beta' rated research applications to funding only a fraction of those in the 'high-alpha' category. The period saw the brain drain turn to a haemorrhage, and a spate of disillusioned early retirals, leaving institutions bereft of some of their best youth and greatest fonts of wisdom. At the time of writing our government seems to persist in the notion that higher education of our youth can be achieved on the cheap, taking no account of the cost of ignorance. The situation is worsened by the market dogma that all research must yield quick returns - ie, be immediately applicable - and the idea that the best direction of research can be decreed by central committees. One wonders how the fundamental researches of Faraday, Maxwell and Rutherford, driven by genius and pure curiosity, but leading eventually to our present day world of electric power, television and atomic energy, would fair in today's grants system.

Appointment to Lectureships in the Department in the late 1960s of Drs D. Clarke and J. C. Brown led to the steady growth of new research areas on top of existing strengths. These were observational spectropolarimetry of very high precision, and diagnostic modelling of high energy processes in solar flares, utilising data from international space missions. The Solar Modelling Group grew out of P. A. Sweet's theoretical interests but also has practical ramifications for earth and space environments, and in turn led on to current interests in image processing, neural network data analysis, and collaborations with the Plasma Physics Group. In the polarimetry work, substantial observing campaigns were conducted here and abroad, and collaboration established with J. C. Brown on interpretation. Research in these fields, together with astrodynamics, produced a growth in the Science and Engineering Research Council (SERC) funded research staff from none to four, with parallel growth in graduate students.

A great deal of departmental research in recent years was influenced by the advent of the space age following Sputnik in 1957. From that time much work was done by A. E. Roy's celestial mechanics team on developing fast numerical methods to compute spacecraft orbits - a need which was especially great in those days when computers were extremely sloth-like and expensive by today's standards. (Around 1968, the Department bought a Frieden 4-function electronic calculator, the size of a television, for £1,000 - ie, £10,000 at 1990 prices - which had a wheel to adjust the decimal place and took about ten seconds to find a square root - a special feature adding £400 to the cost.) NASA's Skylab Missions of the post-Apollo era brought Glasgow's involvement in analysis of astronomical data from spacecraft, including interplanetary dust (D. Clarke) and solar X-rays (J. C. Brown), the latter work evolving through the past 20 years to Glasgow's present involvement with the YohKoh and Compton space observatories. The astronomical polarimetry programme too has used spacebased data, currently from a Wisconsin UV experiment aboard the Space Shuttle. Perhaps most interesting of all is the interdisciplinary space interest which emerged from astrophysics work on radiation pressure effects round stars, which became a celestial mechanics problem, and then a joint project with Aerospace Engineering on solar-sailed spacecraft.

When P. A. Sweet retired in 1982, the Regius Chair was frozen, with many others, as an economy, Professor A. E. Roy being appointed (last as it transpired) Head of the Astronomy Department. At the same time proposals had been floated for amalgamation of the Astronomy and Natural Philosophy departments. These discussions extended over several years, during which the University was persuaded to advertise a new Chair of Astrophysics with the remit of promoting interdepartmental collaboration in research and teaching. This contest was won in 1984 by J. C. Brown, then Reader (aged 37), who, in the same year, had been awarded a D.Sc. for his solar plasma work and elected Fellow of the Royal Society of Edinburgh. Professor Brown was replaced by Dr J. F. L. Simmons, who further broadened departmental research interests into cosmology and statistical studies. In the last year of the separate Astronomy Department, Mrs L. Williamson was replaced on her retirement as secretary by Miss D. Davidson. The departmental merger was finally achieved in 1986, with agreement, currently being implemented, to eventually fund Astronomy accommodation in a refurbished and extended Natural Philosophy building. This departmental merger was the first in a national trend, the UK's last separate Astronomy department (Edinburgh) being about to join Physics.

While the merger marks the formal end of Astronomy's history as a department, it by no means represents an end to the subject. Professor Roy formally retired in 1989, but he remains as active as ever in teaching and research: following the successful conclusion of a consortium (initiated by him in 1983) with Queen Mary and Westfield College London, Liverpool, Royal Greenwich Observatory and Pisa, to study the long term dynamics of the solar system, he has now created a second international research undertaking in celestial mechanics. Such international collaborations are indeed a hallmark of astronomy everywhere, Glasgow having currently joint projects with Sao Paulo, Sydney, Tokyo, Montréal, Madison, Maryland, Berkeley, Alabama, New Hampshire, Copenhagen, Utrecht, Amsterdam, Leiden, Paris, Genoa, Pisa, Rome, Kiev, Budapest and Prague.

After Professor Roy's retiral and Dr Green's half-time appointment as Principal Adviser in the Faculty, the University was (eventually!) persuaded to appoint a new Lecturer in Astronomy and Astrophysics in 1990. The appointee, Dr D. A. Diver, brought extra plasma expertise to the Astronomy Group which, in 1991, merged with Professor E. W. Laing's Plasma Theory Group and became the Astronomy and Astrophysics Group, headed by Professor Brown. In 1991, Dr A. L. MacKinnon, then a SERC Advanced Fellow in Astronomy, was appointed as Lecturer in Physics and Astronomy to the Department of Adult & Continuing Education, and remains an active member of the Astronomy and Astrophysics Group. The Group thus now has more staff, research staff, graduate students, and grant income from many sources than ever in the subject's history and in 1993 will become a SERC funded node in the Starlink computer network. First-year astronomy numbers have recovered from a low of 40 to 80 and a new Arts course ('Exploring the Cosmos') has attracted 40 in its first year. Combined with the astronomy interests of the widely renowned Gravitational Wave Group, led by Professor J. Hough, this puts Glasgow Astronomy in its strongest position ever, looking forward to a healthy future in the Faculty of Science, studying the mysteries of the cosmos.

It is a pleasure to record my thanks to Professor John C. Brown for his help in preparing this account.

Archibald E. Roy is Emeritus Professor in the Department of Physics & Astronomy at the University of Glasgow.

Zoology

Keith Vickerman F.R.S.

Twentieth century biology grew out of 19th century natural history in the great upheaval of thought that followed the publication of Darwin's book *The Origin of Species* in 1859. In this prodigious work Darwin expounded his theory of the evolution of living organisms by natural selection, and gave students of the science of life a new way of looking at both the living and the long dead. Darwin's way was independent of the reductionist approach of the physical scientist - reducing life to physics and chemistry - but just as valid scientifically. This new approach made many biologists recognise that every ecosystem, population, individual or cell, even every biological molecule, has an inescapable past and that biological explanations can be couched in terms of that history. For long, the new approach split biology into two camps, the experimental reductionists (exemplified by physiologists) and the observational evolutionists (latter-day naturalists). Nothing approaching a unified science of life was achieved until the 1930s when the two sides came together in the so-called evolutionary synthesis and Darwinism was reformulated in the language of Mendelian genetics.

Zoologists - students of the science of animal life - in particular have tended to regard themselves as upholders of the Darwinian tradition, and the evolutionist approach has been strong in university departments of Zoology (while the reductionist approach has found more favour in departments of Botany!). By providing an orderly framework for what had previously been a ragbag of unrelated facts, Darwin's theory reduced the enormous factual burden of zoology - a burden so great it was believed to be insupportable to the human mind by the poet Samuel Taylor Coleridge! This burden had in many ways been amplified by Linnaeus in the previous century when he introduced his system of binomial nomenclature and assumed the mantle of Adam in naming animals. The load had increased further with the 19th century craze for natural history and the description of an ever-increasing number of life forms. In this century, evolutionary theory has been increasingly recognised as the backbone of all biology, and today there is as much interest in evolution among molecular biologists as there is among ecologists.

In the University of Glasgow, the development of zoology as a subject has taken place against this background, though zoology in Glasgow and the other ancient universities of Scotland has on the whole maintained a healthy disregard for prevailing fashion. The 'Scottish Professor of Natural History' became a recognisable character to the outside world - eclectic in interests and retaining a need to communicate with the laity long after specialisation and contempt for the amateur had isolated his scientific colleagues in other disciplines. D'Arcy Thompson of Dundee and St Andrews was the supreme example. I have made him a central figure in the account that follows, and I have used changes in academic leadership to reflect the growth of both subject and department.

John Young and the era of comparative anatomy

The Regius Chair of Natural History in Glasgow was founded in 1807 by George III, essentially to provide a curator for the collections of William Hunter. The Professor's remit for Natural History included geology as well as zoology. The early history of the Chair is dealt with by Bernard Leake in this volume, as most of its incumbents were more disposed towards geology than towards zoology, despite the fact that their main job was the teaching of animal anatomy to medical students and that the Chair of Natural History, after standing alone until 1858, had been incorporated into the Medical Faculty. It is appropriate to start this historical narrative, then, with zoology at the end of the 19th century - at the time when the Faculty of Science was instituted - and zoology meant comparative anatomy.

It is not surprising that Napoleon I chose Georges Cuvier, the father of the science of comparative anatomy, to mastermind the reform of education in France. Comparative anatomy - the discipline of the 'correlation of parts' - has many of the virtues associated with classical education in the humanities, development of judgment and a capacity for inference in particular. Cuvier reputedly could infer the form of the collar bone of a bird from a single feather! He sparked off an enthusiasm for comparative anatomy which lasted throughout the 19th century and persisted to dominate the teaching syllabus in many eminent Zoology departments until half way through the present one.

Interestingly, the collections of John Hunter, William's younger brother, played a vital part in this great movement by inspiring its foremost British student, Richard Owen (1804-1892). Owen, who was to become first Director of the British Museum (Natural History), was an Edinburgh graduate who spent his formative years in the Hunterian Museum of the Royal College of Surgeons in London, dissecting countless animals to identify the source of Hunter's material. In addition to discovering the dinosaurs, Owen formulated the concept of homology for organs of similar origin in development. Thus, despite differences in details of their structure, the wings of birds and bats are homologous structures, whereas the wings of insects are simply analagous in function to them. Owen's magnum opus *The Anatomy and Physiology of the Vertebrates* was published in 1866, the year that John Young was appointed to the Regius Chair of Natural History in Glasgow.

In Glasgow, the supremacy of comparative anatomy in the teaching of zoology is abundantly illustrated in the examination papers set by John Young. The demands made on students leave one flabbergasted. Thus in the year of the foundation of the Faculty of Science, science students were requested to "compare an annelid (worm) and a bird as regards (a) segmentation of the body, (b) nervous system, (c) locomotive organs" - all in the space of one quarter of an hour! Medical students were allowed longer for the same question, having to answer four out of six questions in one and a half hours, while science students had no choice and had to answer all six.

The Darwinian revolution had an enormous impact on comparative anatomy, as acceptance of the theory of evolution made homology all the more meaningful. This striking development does not seem to have influenced John Young one bit, however, and in none of his examination papers is the word evolution mentioned. All around him, however, the alliance of comparative anatomy with evolutionary doctrine was refocusing zoology. Demonstration of evolutionary descent was becoming the major preoccupation of zoologists, and comparative anatomists were switching their attention to developmental stages. At the same time, study of the relation of form to function that had fascinated Owen and T. H. Huxley (the giant of academic zoology in Victorian Britain), dwindled in importance. Comparative physiology became increasingly neglected in the late 19th century, to be rescued in Cambridge some three decades later.

John Young did not find the inspiration in William Hunter's anatomical collection that Owen found in John Hunter's hoard. As one of the Commissioners appointed to enquire into the universities of Scotland in 1878, T. H. Huxley complemented Young on his careful scholarship in Greek but expressed regret that he could not find more time for his studies on fossil fishes. Young's delight seems to have been browsing among the books of the Hunterian Library, of which he was custodian. He became an intellectual magpie and ardent Senate man and while his colleagues in University and city benefited from his learning, his students and science at large did not. The distrust of the evolutionary message apparent in Young's teaching did not reflect antagonism to the very idea of evolution on his part, but doubts about the ability of Darwin's theory of natural selection to account for it. These misgivings were expressed in the essays that he penned in his last years in the Chair and which were published posthumously in 1904-5. They do not make easy reading.

Young's scepticism came from two quarters. First, he considered the foundations of Darwin's theory faulty in that the nature of the variations being selected for survival was not understood, and transmission of these variations from one generation to the next was a crucial part of the theory. This reservation is wholly understandable as the discovery of Mendel's work and the advent of the science of genetics had to wait for the new century. Young's second criticism was more emotional: he simply could not bring himself to abandon Paley's argument for a supreme designer and one has to admit that at this time he was not alone in this respect.

John Graham Kerr and dynastic zoology

When, sadly afflicted by gout, John Young relinquished his tenure of the Regius Chair in 1902, his successor, John Graham Kerr, reluctantly saw the name of the Chair changed from Natural History to Zoology and a separate Chair created for Geology.

On graduation from Cambridge, Graham Kerr (he liked to be known by both names) decided to devote himself to the fashionable field of evolutionary embryology. But not for him the commonplace objects of study (fowl, frog, dogfish, rabbit); his fascination lay with 'living fossils' - animals which had apparently come down unchanged from remote geological times. "It seemed to me," he wrote in later years, "that the general principles of vertebrate morphology could not be regarded as firmly established, so long as they ignored the embryology of these archaic types."

Graham Kerr's particular interest was the South American lungfish *Lepidosiren*. The lungfishes or Dipnoi are fascinating relicts of that crucial period in evolutionary history when aquatic vertebrates were beginning to invade the land. An undergraduate expedition to the remote tropical swamps of the Gran Chaco had enabled him to amass one of the most complete series of developmental stages that had ever been obtained for any vertebrate. He could now settle down to several years of research on his prized possessions.

In Glasgow his research went on to show that the simple styliform limb of early vertebrates developed along two lines, one leading to paired fins adapted to swimming, the other to jointed limbs adapted to movement on land. He became a leading exponent of the 'Dynastic School of Zoology' - the 'who evolved from who and how school' which gave comparative anatomy a dynamism it had previously lacked. Collecting the lungfish gave him a passion for South America that never left him, and his book, *A Naturalist in the Gran Chaco,* written in later life, has been described as among the best on that continent since Darwin's *Voyage of the Beadle.*

"The Glasgow professorship with its relatively high emoluments and its long summer holidays was generally regarded by professional zoologists as one of the most attractive in the country" wrote Graham Kerr, later musing on his reasons for wanting the job. He left the well-equipped Zoological Laboratory in Cambridge whose day to day running had been in his charge, to find a vastly different situation in Glasgow. Here there was no zoological laboratory to speak of. The eastern half of the lower hall of the Hunterian Museum - known in later years as the Hunter Hall - had been used for what little practical work had been done, but it had no scientific equipment, he noted. Its floor was littered with geological debris. Its entry was by wooden steps through one of the windows. The renowned Hunterian Collections were in a parlous state and he described how he found the famous collection of insects, worked through and arranged by the eminent Dutch entomologist Fabricius, in a decrepit chest, discarded outside in the Quadrangle.

Scientific equipment to Graham Kerr meant microscopes. The bright field light microscope had reached its zenith in Germany towards the end of the century, and microscopical studies were a necessary part of embryology. In teaching, he took an obvious delight in seeing the ecstatic response of students peering at living protozoa down the microscope for the first time. The development as teaching aids of the museum on the one hand, and microscopy on the other, plus rehousing of the entire Department became his main missions in life.

In 1923, 20 years after his arrival, these ambitions culminated in the opening of a monumental new building for Zoology, the one that houses the Department (or most of it!) today. This building is undoubtedly one of the finest in the University. The late Professor Carl Browning F.R.S. told me that Graham Kerr's inspiration for its style was the Baths of Caracalla in Rome! If this was true, his ideals must have been tempered considerably by the architect, Sir John Burnet. Originally the building had two notable features, the Museum and the teaching laboratory. The former, with its imposing columns surrounding a central floor well, certainly conjures up visions of languorous figures in togas awaiting a dip, while the laboratory was described by one of Graham Kerr's successors (Sir Maurice Yonge) as "the most magnificent in Great Britain, perhaps in Europe". It was designed to accommodate 150 students. Not only did it have an impressive floor area, it rose the full height of the building, its extensive north facing windows flooding the benches with light (on such days in Glasgow when light was available). A 'good northern light' was believed by Graham Kerr to be essential for dissection and microscopy.

In order to develop the Museum for teaching, he bartered some of his precious dipnoan and archaic fish material for other exotic specimens with dealers. He was particularly proud of the gorilla skeleton obtained from Paris at the outbreak of World War I; it had been intended for Germany. He obtained parasitic animals from his former medical students, many of whom went on to specialise in tropical medicine.

Medical students formed the greater part of Graham Kerr's lecture class. In teaching he abandoned the T. H. Huxley tradition of starting with dissection of the rabbit, preferring to begin with animals of 'the least possible complexity' - ie, Protozoa. The obligatory survey of the animal kingdom he confined to members that illustrate important principles and those of particular importance to medical science. He continued to be a much-appreciated lecturer in the Department for over 30 years and he thrived on student response to his teaching. His 'Daft Friday' lecture became an institution and recounted with great gusto his trips to the Gran Chaco. It is remembered with affection by all who heard it - even on those occasions when it lasted over three hours.

As an undergraduate Graham Kerr had visited German universities including the Department of the great evolutionist Ernst Haeckel in Jena. Though his stay was brief he had absorbed enough of the scientific team spirit to become captivated by it and wanted to see his Glasgow Department modelled on similar lines. He went to great lengths to secure a talented histology technician named Jamieson from Edinburgh as he realised that his anatomical work needed such assistance if he was to compete with the German embryologists. The value he placed on this assistance is reflected in the story of Graham Kerr's election to the Royal Society in 1909 - a legend so famous I have heard it repeated in several after dinner speeches, and on at least three occasions ascribed to different zoologists. The election happened to coincide with the birth of his twin sons. His rumbustuous medical class had been told of the latter by their mothers who had read it in the *Glasgow Herald*, and they greeted his entry to the lecture theatre next day by finishing their usual sing-song with a tumult of cheers. They knew nothing of his F.R.S., but he presumed that this was what the noise was about. With great solemnity he confessed that he could never have achieved this cherished ambition without the help of his trusted histology technician!

Although very much a vertebrate man, Graham Kerr's obsession with microscopy and his love of protozoa for introducing the animal world to students, led him to encourage protozoology in the Department, and to launch the careers of three remarkable women. He was a vigorous promoter of women students, having taught zoology to students of Girton and Newnham Colleges when his superior at Cambridge, Professor Adam Sedgwick, refused to do so. In Glasgow he insisted on teaching women alongside men instead of separately in Queen Margaret College as was the custom. Zoology was long considered a somewhat indecent interest for a woman. As late as 1944, the appontment of Doris Mackinnon, the first woman professor of Zoology in the UK, to the Chair of Zoology at King's College, London, was greeted with howls of protest from the theological faculty who were outraged at the thought of her dissecting a male mammal in front of students.

Few women students can have endured the difficulties encountered by Sister Monica Taylor, for many years head of biology at Notre Dame College, Dowanhill. Although she attended Graham Kerr's classes in the early 1900s as a Roman Catholic she was not allowed to graduate with a Glasgow degree and had to take a London University external degree instead.* She developed large scale farming of *Amoeba proteus*, the mainstay of Graham Kerr's introductory practicals, and isolated in culture, from Loch Katrine and elsewhere, amoebae now used in cell biology research throughout the world. In her nineties, the University relented and awarded her an honorary LL.D. Her interest in amoebae was shared with Dr Margaret Jepps who from 1923 to 1952 represented protozoology in the Department. She was an expert on parasitic amoebae and described a new parasite of the human gut, *Dientamoeba fragilis*. Two recipients of her lectures, Professors A. J. Haddow and W. H. R. Lumsden recorded their admiration for her teaching in a famous valentine card, decorously adorned with protozoa, and recently exhibited by the Natural History Museum.

* Her trials were not all academic. In the Second World War, she was arrested on suspicion of being a German parachutist disguised as a nun! When asked to prove she was a woman, she simply pointed at her feet!

One of Graham Kerr's most outstanding students was Muriel Robertson, the first woman zoologist to be elected F.R.S. (in 1947) and one of the first women to be elected to the Royal Society. After graduating in 1908, she pursued research on fish trypanosomes in the Department. She later joined the Royal Society's Sleeping Sickness Commission in Uganda, investigating the nature of this serious disease then sweeping up the eastern side of Africa. She worked out the cycle of development of the sleeping sickness trypanosome in its tsetse fly vector and thus made a lasting contribution to tropical medicine. My own research on trypanosomes essentially carried on from hers 50 years later. Diminutive in stature, she remained a formidable figure in protozoology into her ninth decade when I came to know her. In her handbag she carried a grenade to brandish at motorists who harassed her on zebra crossings.

In 1936 Graham Kerr resigned from the Regius Chair to enter Parliament as Unionist MP for the Scottish universities, succeeding the novelist John Buchan on his elevation to the House of Lords. He defeated his Socialist adversary, the writer Naomi Mitchison, by over 16,000 votes. On leaving Glasgow he made over his scientific library, instruments and zoological collections to the Department.

In many ways he was not an obvious parliamentarian, declaring himself "profoundly suspicious of debate as a method of reaching sound judgment" and well aware that "actual knowledge has a paralysing effect on eloquence". He had originally joined the Unionist Party in 1924 as a protest against the appointment of the first Socialist Prime Minister, Ramsay MacDonald, and rose to become President of the Scottish Unionist Association. In Parliament he took a particular interest in education, fisheries and, in the Second World War, in the camouflage of battleships. This last was an old hobby horse of his, and grew out of an interest in animal colouration derived from his tropical experience.

Inspired by the work of the American G. H. Thayer on animal camouflage, he had written to Winston Churchill, then First Lord of the Admiralty, in 1914, suggesting ways of making warships less conspicuous through obliterative shading and outline disruption. In the Second World War this method of protection was adopted almost universally, even for land installations. In 1933 he had invited to the University as lecturer and advisor on these matters, the renowned photographer and artist-zoologist, Dr Hugh B. Cott, whose special interest was mimicry and animal camouflage. A photograph in the Zoology Department shows the main lecture theatre set out for one of his lectures. The blackboard is covered with magnificently-executed drawings and half the contents of the Museum seem to occupy the demonstration bench where they had been laid out by three technicians. His book *Adaptive Colouration in Animals,* published in 1940, was to become a classic in the naturalist's library. Cott's appointment, like that of G. S. Carter before him, was significant in that it showed recognition of field studies as a serious component of zoology after decades of consignment to an academic dustbin by zoologists obsessed by morphology.

John Graham Kerr was knighted in 1939, long after he had left the University of Glasgow. His ex-students, however, invariably referred to him as "Sir John", though in a Freudian slip in her recorded memoir of him, Sister Monica Taylor refers to him as "Saint John"! Today the portrait of this canonised figure in the Department's history regards us with a critical eye from the head of the main staircase in the building that stands to his memory.

Edward Hindle and medical zoology

As a staunch upholder of the dynastic school of zoology, Graham Kerr had cold-shouldered experimental work, but spear-headed by the School of Zoology in Cambridge, experimental biology was becoming all the rage in the 1930s. The Society for Experimental Biology had been founded in 1925. Ironically for Graham Kerr, the 1930s were the hey-day of experimental embryology, with attempts being made to test Spemann's theory that differentiation in the developing embryo is the outcome of an orderly sequence of inductive stimuli. As previously remarked upon, the '30s also saw a major coming together of reductionist and evolutionist thought in the Evolutionary Synthesis, but Glasgow's new Professor of Zoology was not an obvious exponent of any of this.

In succeeding Graham Kerr in the Regius Chair in 1935, Edward Hindle created an upheaval in the traditional zoology courses. He introduced a more applied bias, orientated towards entomology, as among his many interests, the tranmission of disease by arthropods figured strongly. After training with G. H. F. Nuttall in parasitology at Cambridge, he had been Professor of Biology in Cairo, but returned to full time research at the London School of Hygiene and Tropical Medicine. Here he conducted investigations on tick transmitted diseases but left to join the Royal Society's Kala Azar Commission in China studying transmission of the disease by sandflies; Kala Azar is caused by the protozoan parasite *Leishmania donovani,* named after Sir William Leishman, a Glasgow graduate who identified the parasite in the spleen of Kala Azar patients. By transmission experiments, Hindle demonstrated an important principle of vector biology - that local insects are more susceptible to local strains of the parasite than to strains from elsewhere. On returning to London and the laboratories of the Wellcome Foundation, Hindle switched to arthropod-borne viruses and immunology, and he made notable contributions to the development of one of the most successful vaccines ever devised - that against yellow fever. He is perhaps remembered best, however, as the saviour and populariser of the golden hamster in research. From his original two pairs, all the golden hamsters of Europe and elsewhere are derived.

In his eight years in Glasgow, Hindle did relatively little research, but did much to popularise zoology. He founded the Zoological Society of Glasgow, with a view to instituting a zoological garden in the city and acquired Calder Park for this purpose. The outbreak of war upset plans for its development, however, and the Zoo did not open until 1946, long after Hindle had forsaken the Chair to become Scientific Director of the Zoological Society of London in 1943. He had a somewhat military bearing and liked being in charge. In 1942 he became commanding officer of the University Officers' Training Corps and of the 14th Battalion City of Glasgow Home Guard. He followed up his Presidency of the Royal Philosophical Society of Glasgow and Chairmanship of the Scottish Marine Biological Association with presidencies of the Universities' Federation for Animal Welfare, the British Association for the Advancement of Science and the International Wildfowl Bureau. He also became Honorary Secretary of the Royal Geographical Society. Most important of all, perhaps, he was founder and first President of the Institute of Biology. He was elected F.R.S. in 1942.

One of Graham Kerr's last graduates, J. D. Robertson, returned with a Ph.D. from Cambridge in 1939, as Lecturer and later Titular Professor in the Department. He brought experimental comparative physiology to the Department and with his studies on ionic regulation in marine animals founded a tradition in that subject that continues today. Another notable appointment was Otto Loewenstein who founded another of the Departments's traditions, neurophysiology; he left in 1952 to take the Zoology Chair at Birmingham. What was to be virtually Hindle's last scientific paper, on meiosis in the body louse, was written with Guido Pontecorvo whom Hindle had invited to the Zoology Department in 1942 and who was to become first Professor of Genetics in the University.

Hindle's technician, Mr J. Smith, continued to visit the Department into the early 1980s. He recalled how his duties included putting up shelves and laying carpets in Hindle's house in University Avenue; in return the Professor would take him shopping in his Rolls Royce! A curious relic of Hindle's stay in Glasgow is the frieze of animals adorning the south wall of the Chemistry building. It was allegedly put there to assuage the apoplectic incumbent of the Regius Chair of Zoology who was furious at having his light and vista obstructed by a yellow brick building of decidedly inferior architecture to his own.

C. M. Yonge and marine biology

In appointing Charles Maurice Yonge to the Regius Chair in 1944, the University of Glasgow acquired a most distinguised zoologist. His early pioneering work on marine invertebrate feeding and digestion led to his appointment as leader of the highly successful Great Barrier Reef expedition of 1928-29. In 1932 he had become first Professor of Zoology at the University of Bristol where he remained until his move to Glasgow. His research interests were broadly within marine biology but his major contributions to science were on the symbiosis of marine animals with algae and on the relation of form to function and evolution in the bivalve molluscs. He was a dedicated scientist in both the reductionist and evolutionist traditions, and a scholar with wide interests, especially in history. He told Roy Plomley on the radio programme *Desert Island Discs* that he was probably the only scientist to read Gibbon's *Decline and Fall of the Roman Empire* sitting on a coral reef! He was also an inveterate traveller, a fine teacher and a prolific writer of eminently readable prose. His books on *The Seas* (with F. S. Russell), *The Sea Shore* and *Oysters* are classics, and inspired many young zoologists, including the present writer. He was elected F.R.S in 1946 and knighted in 1967.

If Graham Kerr had been shocked by the physical state of the Zoology Department that he inherited, Maurice Yonge was by turn impressed, daunted and amused with what he found. Although Graham Kerr's building was imposing, it was designed "as though the needs of 1923 were those of eternity" and he saw a long struggle ahead adapting it to the demands of Zoology in the second half of the 20th century. He was amused because his predecessor had, under wartime exigency, converted the place into an urban farm, with livestock in every conceivable corner.

He inherited an academic staff of four. Charles Parsons (a vertebrate zoologist whose *Nature* paper describing the consumption of an alligator by a mere toad must be one of the most droll ever to appear in that journal), Otto Loewenstein, Margaret Jepps and Agnes Miller (a protegee of Graham Kerr's and another devotee of arachaic fish). He found only two Honours students - L. H. Finlayson and W. Russell Hunter who were to go on to occupy

Chairs in Birmingham and Syracuse in New York respectively. But with the end of hostilities, staff returned from wartime occupations and student numbers began to climb rapidly. It was necessary to adapt the Department physically, and during his tenure of the Chair almost every part of the original building was modified. The first building erected by the University after the War was a western extension ('The Annexe') to accommodate agricultural zoology. Opening up the wall of the museum involved cutting through three feet of solid masonry (it is somewhat ironic that at the time of writing this opening is being filled in again and the annexe demolished to make way for the new central animal facility!). Similar problems were encountered in reconstructing Graham Kerr's pride and joy, the elementary laboratory, by dividing its vast space both horizontally and vertically to provide several laboratories and a small lecture theatre. One workman resigned because he felt defeated by the thickness of the walls. A later development was peripheral subdivision of the pillared hall of the Museum to create a further lecture theatre and storage rooms. The catacomb-like basement was opened up at the same time that the annexe was built.

Agricultural zoology represented one of the many facets of entomology studied in the Department over long periods of its history, though sadly now very under-represented in the Department's staff. As they form almost three-quarters of all described living species (with many, many more still to be described), insects have to be important and it surprises me that most laymen do not consider them animals! The Department has long provided an identification service for insects sent in by the public. Starting after the War, Dr Alex Hill, with his colleagues Alastair Fraser, Ron Dobson and Douglas Cochrane taught a highly successful degree course in Agricultural Zoology with a strong entomological component in the Department until lack of full-time expertise made it necessary to discontinue it in the late 1980s. Also in the late 1940s, the UGC offered the University a Lectureship in Taxonomy to which Dr Roy Crowson was appointed in 1948. Crowson became one of the world's leading authorities on insect systematics, especially as applied to the Coleoptera (beetles) which constitute one quarter of living animals. Dr Kenneth Lockey has explored the use of cuticular lipids in insect taxonomy in more recent years. In the last 20-30 years, taxonomy has been neglected, even discouraged, by the research councils but amends are now being made and the University has just become the recipient of a Natural Environment Research Council award for the redevelopment of taxonomy in the UK before existing expertise disappears.

Other notable entomological work in the Department was of an ecological nature. Dr M. V. Brian published a series of valuable papers on ant populations between 1946 and his departure to take charge of the Nature Conservancy research station at Furzebrook in 1953. Later Dr A. F. G. Dixon developed terrestrial ecology with a study of population dynamics of the sycamore aphid as a preliminary to an analysis of energy flow in woodland communities.

The development of Loch Lomond as a field base for freshwater biological (limnological) research was also started in Yonge's time. Two large ex-army wooden huts erected on the property of Sir Iain Colquhoun at Luss formed the first University of Glasgow Field Station. It accommodated the work of Dr Harry Slack, who had instituted a field course in limnology in the Department before leaving on military service), and Dr Anthony Downes who, with Drs Douglas Kettle and Bill Lawson, had set up a programme of research on the species and life histories of the biting midges (*Culicoides* spp.) for which Scotland is notorious. Slack's extraordinary technical and inventive abilities enabled him to manufacture most of the equipment needed from military surplus supplies. After 1964 his studies and those of Dr Peter Maitland on bottom deposit fauna and the fish of Loch Lomond continued in more purpose-built accommodation across the loch on a site within the Queen Elizabeth Forest Park near Rowardennan. Official opening of the University Field Station took place the year after Maurice Yonge's retiral.

The take-over of the Glasgow Veterinary College by the University resulted in the transfer of Dr (later Professor) C. A. Hopkins to the Zoology staff. When the Institute of Biology's Parasitology Group (later the British Society for Parasitology) was founded in 1957, Adrian Hopkins was undoubtedly one of its liveliest wires and while pursuing research first on tapeworm physiology in relation to the gut of the host and later on immunity to tapeworms he set up the Department's reputation for helminthological parasitology. A grant from the Wellcome Trust enabled him to move his research operations to the Wellcome Laboratories of Experimental Parasitology attached to the Veterinary School at Garscube in 1963. I shall defer discussion of parasitology in the Department until later.

When Charles Parsons died tragically and suddenly in 1950, he was succeeded as Lecturer in charge of the medical class by Anthony Barnett, who was to develop the study of animal behaviour in the Department. One of the greatest revolutions in zoology this century concerned the way in which animal behaviour is studied. In the 1930s, the mania for experimentation led to the belief that the only way to investigate behaviour scientifically was through some kind of experimental intervention; simply to observe animals was what amateur bird watchers did. The Nobel Laureates Konrad Lorenz and Niko Tinbergen who founded the new science of ethology in the 1940s did precisely this, however. They studied natural rather than contrived behaviour, and were able to discern natural behaviour episodes. They employed the comparative approach of the evolutionist and demonstrated the occurrence of the same or similar behavioural sequences in members of related species as though they represented the playing out of a genetic programme, and they were concerned with the adaptive significance of behaviour. It was not informative to study variations of behaviour unless we already knew the norm from which variants depart, they said. Barnett's extensive studies of wild rat behaviour in captivity and his book, *Modern Ethology*, published after he left Glasgow for the Chair of Zoology at Canberra in 1971, showed that while having one foot in the ethologist camp, his other was in the reductionist behaviourist camp that preceded ethology.

Neurophysiology, the subject of Otto Loewenstein's research, was continued after his departure by Dr Graham Hoyle who concentrated on neuromuscular transmission studies with insects. When Hoyle took the Chair of Zoology at Oregon, his junior colleague Dr Peter Usherwood carried on this research and played a leading part in the search for non-cholinergic transmission mechanisms in the insect nervous system.

The Second World War had led to rapid expansion of interest in the sea and in marine biology; maximising the use of natural resources, combating natural hazards such as the fouling of ships by marine organisms, and understanding factors that determine water clarity were just some of the more obvious problems being faced up to by marine biologists. New techniques developed from defence work could now be applied to develop understanding of processes in the sea. Marine biology was set for a growth spurt and the University of Glasgow's Zoology Department under Maurice Yonge could contribute to it. Yonge's particular interest in invertebrate feeding was of obvious relevance to marine productivity. The appointments of my present day colleagues Peter Meadows (interested in the effect of animals on marine sediment stability) and Dr Peter Spencer Davies (studying coral biology) date from Yonge's time.

Yonge served on the Advisory Committee on Fishery Research to the Development Commission from 1937 to 1956. This committee coordinated marine research in the UK; its function was taken over in 1965 by the Natural Environment Research Council and Yonge became one of its first members. He also served on the Colonial Fisheries Advisory Committee (later an advisory panel to the Ministry of Overseas Development), and this led him to visit many parts of the world where fisheries were being developed. These contacts also enabled him to fix up sabbaticals (in which he was a great believer!) for several of his colleagues and broaden their field and laboratory experience. He himself had a series of sabbatical absences - first as Visiting Professor at Berkeley, California in 1949, then at the University of Washington Seattle, 1959-60, plus a spell in Jamaica and the USA in 1964 - a modern professor drools with envy..!

Yonge's prodigious output of original research papers, reviews and books on marine biology and molluscs was aided by a fine supporting staff and the proximity of the Scottish Marine Biological Association's laboratory at Millport, where all marine activities of the Department took place in his time. Like his two predecessors and his successor in the Chair, he was, from 1944 to 1967, Chairman of the Executive Committee and then President of the Scottish Marine Biological Association and closely concerned with the Millport Laboratory. Yet he regarded Millport as a most unsuitable place for the Association's research laboratory and devoted much effort to getting the laboratory moved (eventually, in 1966) to its present site at Dunstaffnage near Oban, where proximity to the deep sea and continental shelf opened up a whole range of novel opportunities in marine research.

Sharing in his bivalve interests were a series of eminent protegees - Gareth Owen, later to become Professor of Zoology at Belfast and Principal of University College, Aberystwyth, John Allen, later Director of the Millport Marine Station after its University take over, Russell Hunter, and Alan Ansell. His personal research, he acknowledged, owed much to Dr Hugh Steedman, who had followed him from Bristol to Glasgow, and developed technical services in the Department, especially histological, photographic and workshop facilities, as well as performing much routine administration on his behalf. Jan Campbell (still with us overseeing first year biology classes) helped him keep research going during periods of heavy administrative commitment, as when he became Dean of Science from 1960-63. This last commitment seems to have made him tire of being a professor and he sought permission to retire from the Chair early, in 1964, in order to devote more time to research and travel. He assumed the status of Honorary Research Fellow in the Department and later at his *alma mater,* the University of Edinburgh.

Retiral from the Chair not only enabled him to get on with his herculean task of surveying the bivalves, it freed him for participation in several Royal

Society committees - the Southern Zone Research Committee (1966-79), the Aldabra Research Committee setting up the Royal Society's Field Station on the island, the Pacific Science Association Committee and so on. He also served on the Nature Conservancy Council. He was awarded the Darwin Medal of the Royal Society in 1968. He received honorary degrees from Bristol, Heriot-Watt, Manchester and Edinburgh universities and innumerable honorary memberships of learned societies. He can be counted among the truly great naturalists of this century.

David Newth and sixties expansion

In 1965 the challenge of developing the now well-established Zoology Department fell to David Richmond Newth on his appointment to the Regius Chair. David Newth's early training and first appointment at University College London (UCL) had given him wide experience in both classical and experimental zoology, and developmental biology in particular. His most important research was on the development of the neural crest in amphibians. In the 1950s he was one of a brilliant circle of biologists that centred around the eminent geneticist J. B. S. Haldane and the Nobel Laureate P. B. Medawar in the Zoology Department at UCL. All had a lively interest in basic philosophical problems of biology, its public face and its social relevance. In David Newth's case these issues increasingly took precendence over bench research.

At the time of his Glasgow appointment, Newth was a well known figure in his own field of experimental embryology. Together with the late Michael Abercrombie F.R.S., he had founded the *Journal of Embryology and Experimental Morphology* (now *Development*) in 1953 and was to remain its editor until 1969. Under the auspices of his editorial board, an extremely influential series of international embryological conferences was inaugurated. He was a prime mover in the evolution of the London-based Embryologists' Club into the now flourishing British Society for Developmental Biology.

The elucidation of the structure of DNA and confirmation of its role in heredity in the 1950s launched the era of molecular biology and brought genetics to centre stage. In 1960 when Newth followed the philosopher J. H. Woodger as Professor of Biology as applied to medicine at the Middlesex Hospital Medical School, he was among the first to attempt genetic transformation of higher animals (axolotls to be precise) with extracted DNA. Such transformation is routine today and forms the basis of gene therapy, but Newth's experiments were not successful and on moving to Glasgow he was content to put his own research interests in second place to transforming teaching and research in the Zoology Department.

Under his direction the academic advantages of the Department were exploited to the full. In teaching, he introduced biology courses to replace the existing separate botany and zoology courses for first year students in Science and Medicine in recognition of the unification movement taking place throughout the country. Later he started a new human biology course for Arts-based students, in fulfilment of a long-held belief that Arts-educated teachers and administrators would benefit no end from it. He also persuaded his colleagues to lay on a similar course for Jordanhill education students. His major innovation in the Zoology Honours course was to bring in options in third year, to allow a greater element of student choice in the curriculum and more time for teaching as well as greater depth in certain fields. This move reflected his recognition that zoology as a science was breaking up into more specialist disciplines. With the encouragement of Professor Alexander Haddow F.R.S., Administrative Dean of Medicine and one of the most distinguished graduates of the Department, this system enabled the Department to launch new intercalated degrees in Parasitology and Animal Developmental Biology to add to the Department's existing repertoire of Zoology and Agricultural Zoology.

That all these new courses prospered was in large measure due to David Newth's keen interest in teaching at all levels and his genuine care for students (plus, one has to say, a certain amount of forbearance on the part of colleagues who gave up research time to take on more teaching!). An engrossing lecturer himself, he introduced a rigorous system of checking on course efficiency and student response which long antedated quality assurance. A striking growth in student numbers - quadrupling of the Higher Ordinary Zoology and Honours classes, and doubling of postgraduate numbers - accompanied these changes. Graham Kerr's building was bursting at the seams and constant conversion was necessary within it, despite the removal of first year students to the new Boyd Orr Building in 1970 and withdrawal of the elementary class for veterinary students.

Mercifully in the early '70s, increased student numbers led to a compensatory increase in numbers of staff to teach them! The academic staff grew in number from 18 to 34 on a curious 'Noah's Ark' basis, except that while Noah found two of each kind sufficient for his ark, Newth aimed at three teachers in each subdiscipline (ecology, developmental biology, protozoology and so on).

Diversification came with increasing staff numbers and research groups proliferated.

David Newth both fostered research groups within the Department and imported them from elsewhere. A flourishing group had been built up around P. N. R. Usherwood studying the simple nervous systems of insects and the remaining unconverted side galleries of the museum were carved off to make a large laboratory to house them. The ecologist A. F. G. Dixon set up another research group on aphid biology, accommodated by creating research laboratories for them on the first floor. The success of these enterprises was recognised when their leaders were appointed to Chairs in Nottingham and East Anglia respectively in the early '70s. Dr Walter Edgar working on spider ecology and Drs Jon Barnes, Martin Burns and Douglas Neil, all neurophysiologists, helped to fill the gaps left by these departures. Dr Mike Hansell was left to carry on ethological studies with his work on animal architecture after the departure of Anthony Barnett.

Funds from the Nuffield Foundation enabled me to move from University College London to Glasgow in 1968 and set up a research unit in Protozoology, studying structure-metabolism relationships in trypanosomes. Drs Rob Bruce, Stephen Phillips and Derek Wakelin were recruited to add expertise in the growing field of immunological parasitology, and Dr Graham Coombs to develop interests in parasite metabolism and drug design. Dr Ann Lackie filled the unusual niche of invertebrate immunologist with parasitological leanings. The SERC, AFRC and Leverhulme Trust brought Dr Donald Ede from Edinburgh to continue his work on vertebrate limb development and found a new Developmental Biology Unit. The Department was extended to a fifth site when the latter group, joined by our other developmental biologists Drs Suzanne Ullmann and Roger Downie, was installed at Horselethill.

Glasgow Zoology was introduced to life history studies by Dr Peter Calow who joined us as physiological ecologist, bringing an evolutionary approach to the physiology of reproduction, development and ageing. He soon had the Department reverberating to the language of the new ecology with its metaphors taken from the battlefield ('strategies') and the stock exchange ('investment'). The Field Station at Rowardennan, after Harry Slack's retiral under the direction of Dr Roger Tippett, was meanwhile, with the help of Stirling University, extended to provide additional laboratories for this work and extra accommodation for students. On the marine side Dr Alan Taylor brought additional expertise to physiological ecology, working with collaborators at

Millport, now associated with the universities of London and Glasgow after the transfer of the SMBA to Dunstaffnage.

With other new appointments, the Department became quite a centre for applied ornithology, both ecological and behavioural. The behaviour of sea birds and their role in transmission of disease was Dr Pat Monaghan's principal interest, while Dr Bob Furness concentrated on sea bird ecology in relation to fisheries and the mounting problem of pollution. Dr David Houston's work on the ecology of vultures and other scavenging birds brought our students a welcome contact with the tropical world. Dr Felicity Huntingford introduced behavioural studies on fish, latterly with emphasis on the salmon. All these activities are still ongoing. Dr Peter Norton, who was with us as a Lecturer in the '60s and '70s before leaving to take holy orders was the last representative of the world of palaeontology and dynastic zoology which earlier in the century had figured so strongly in our subject.

With this increase in number of academic staff and technicians nearing 40 in number, the Department became one of the largest Zoology departments in Europe and in academic burden (in terms of student FTEs) the fourth largest department in the University (Graham Kerr, before the move to his new building, used to entertain the entire Department, staff and students, to tea in his rooms!). Generous help was given in teaching by the staff of the Universities Marine Biological Station, Millport, the SMBA laboratory at Dunstaffnage, and in particular the Beatson Institute for Cancer Research under the direction of Dr John Paul, whose staff provided the necessary molecular component of the Developmental Biology degree. The Department of Immunology, and especially Dr W. J. Herbert provided much appreciated courses in immunology for our students.

In the late '70s, and ostensibly in response to the mounting pressure of departmental administration, the Senate agreed to divorce the administrative role of departmental head from the appointed Chairs. In 1979 David Newth took advantage of the change to demit office as Head of Zoology and assume the Deanship of the Faculty. He was a consummate administrator and much in demand for committee work. He was currently Chairman of the Nature Conservancy Council's Advisory Committee on Science at a critical time when the Wildlife and Countryside Act was going through Parliament, and his contributions as a scientist were very much appreciated by other Council members. In 1981 he announced that he was retiring to the wilds of Argyll and spent his last seven years in isolation among the hills and the golden eagles. His urbanity and civilised conversation are still missed in the Department.

New Chair, new era

The University Court made provision for a second Chair in Zoology in 1965 but not until 1979 was the Chair filled when I was appointed to it and simultaneously became Head of Department. Although the Department expressed a wish that the Chair should be named in honour of Sir John Graham Kerr, Court did not get around to naming it until 1984 when I was translated to the Regius Chair and David W. T. Crompton from Cambridge University was appointed the first John Graham Kerr Professor of Zoology in 1985.

Lean times were upon us! As soon as I had been appointed to the Chair, I was told that my replacement lectureship would not materialise - and worse was to follow. A vigorous early retirement drive by the University led shortly afterwards to a massive efflux of staff that with loss of others to appointments elsewhere reduced our academic establishment from 35 to 22. My main job seemed to consist of delivering eulogies at farewell parties. After reluctantly approving the early retiral of so many valued colleagues I learned that there could be no replacements for them either.

The problems facing the Department were twofold: first, how to cope with a massive teaching programme with only two-thirds of our former staff and, secondly, how to survive as a research department. The research councils were insisting more and more on focusing of research strengths - and we had prided ourselves on our diversity! Research was now acknowledged to be a team activity and the Noah's Ark system of appointments in Zoology had encouraged the appointment of individuals who would pursue their own interests irrespective of others. My predecessor had regarded the overgrowth of any particular aspect of zoology as undesirable. But in my own area of parasitology, Adrian Hopkins and I had a strong say and managed to ensure that colleagues whose interests complemented our own were appointed. The University had a good reputation in parasitology and it was clearly a subject to promote in the Department's interest. It was also a subject with a changed reputation.

I had entered parasitology at an inauspicious time. In the mid-'50s, the subject had been too-long immured in the schools of tropical medicine and seemed isolated from (if not impenetrably deaf to) developments in the rest of biology. Indeed, while an undergraduate at UCL, my mentor, Professor P. B. Medawar, had warned me off it, describing it as a somewhat philistine pursuit!

Protozoology, however, he assured me, was much more intellectually respectable; he was impressed by the work of the Sonneborne school who had recently shown that in protozoa there is more to heredity than the chromosomes in the nucleus and that the cytoplasm has its own genes. What is more, he told me, there is a worrying lack of protozoologists in the UK. I was convinced: I would become a protozoologist. But my other great mentor, the geneticist J. B. S. Haldane had persuaded me that parasites could play an all important role in promoting genetic diversity in evolution, so I stayed faithful and became a parasitic protozoologist. Medawar was not too upset for, after three years absence, he invited me back to UCL where I taught parasitology (and the rest of zoology!) for ten years before moving to Glasgow.

Crucial to my own development as a biologist had been the profound change in our understanding of the physical basis of life, and the substitution of the structural for the colloidal concept of the contents of the living cell. In mid century, the electron microscope had shown that there is no such stuff as 'protoplasm' and there is no dividing line between structures in the molecular and in the anatomical sense. This was one of the most important revolutions of biology. I was privileged to be among the first UK biologists to use one of these marvellous machines, under the tutelage of Dr David Robertson, inventor of the concept of the 'unit membrane', at UCL. The excitement of exploring uncharted regions of the cell was unforgettable. I like to think that I played a part in opening up of parasitology to the outside biological world in the early '60s when advances in cell biology, immunology and later molecular biology were proceeding apace.

On coming to Glasgow, one of my first concerns had been improvement and expansion of the electron microscope laboratory. With the acquisition of Laurence Tetley as a talented technologist in 1974 this latter facility has gone from strength to strength and now services all the biological sciences from the new EM Centre in the Chemistry building.

In attempting to build up parasitology at the dawn of the '80s, the loss of Adrian Hopkins through early retirement and Derek Wakelin to a Chair in Nottingham were bitter blows. In the absence of any prospect of replacements, the Wellcome Trust kindly gave us one of their University Awards for a helminthologist and Dr Malcolm Kennedy was appointed. He joined Stephen Phillips in moving into the vacant Wellcome Laboratories. But with all my senior colleagues gone I was sadly bereft of peer advice and it was a great relief when the University Court agreed to filling of the vacant Chair on the departure of Peter Calow to the Zoology Chair in Sheffield University.

With David Crompton's appointment as John Graham Kerr Professor and Stephen Phillips promoted to a Titular Professorship we now had a plural professoriate again. David Crompton's broad experience in parasitology, especially in helminth physiology in relation to nutrition of the host, was a major acquisition and earned the Department recognition as a World Health Organisation Collaborating Centre in 1989. This provided a much-needed boost to our ambitions, especially as David Barry from my own group had been transplanted to the new Wellcome Unit of Molecular Parasitology created for Andrew Tait (attracted from Edinburgh and also an Honorary Lecturer in Zoology here) and himself and housed in the departments of Veterinary Parasitology and Genetics. Glasgow was certainly getting a name for itself in parasitology!

In 1985 Stephen Phillips succeeded me as Head of Department. He had a hard cross to bear when in February 1988 the roof laboratories, built to accommodate comparative physiology in 1965, were consumed by fire and many of our colleagues consigned to portakabins outside the main Department. It took three years of delicate negotiations to get the agreement of the University and city authorities that a replacement building should be constructed on the roof of the museum, but the end result was worth waiting for and was opened by the Secretary of State for Scotland, the Rt. Hon. Ian Lang M.P. in June 1991. The new laboratories are mainly occupied by David Crompton's research group, a new helminthologist Lecturer, Dr Paul Hagan who works on schistosomiasis, and the developmental biologists who moved back onto Gilmorehill following the retiral of Donald Ede.

The surrender of the Horselethill Developmental Biology Laboratory entitled the Department to extra space on the campus precinct and this was provided in the reorganised Chemistry building. Just before the opening of the new roof laboratories, Graham Coombs, recently promoted to Titular Professor and more recently to Head of Department, and I acquired splendid new laboratories in the basement area. This laboratory for biochemical parasitology is already bursting at the seams. Meanwhile the demolition of the annexe to make way for the new central animal house facility has consigned more colleagues to the portakabins which look as though they are going to be around for some time.

It would be misleading to imply that recent development of the Department has been confined to parasitology. Despite the shortage of new appointments, new ideas have transformed the outlook of our colleagues devoted to other aspects of zoology. This is especially true for those concerned with animal behaviour and ecology who are forging a common interest in life history strategies. As I discussed earlier, for much of this century zoology was largely concerned with the dynastic evolution of animals and the demonstration of the central role of morphological and physiological adaptation in this process. Reductionist ecologists in the post-war period have been mainly preoccupied with energy flow through ecosystems and the recognition of food webs. In more recent years evolutionary ecology has come to the fore. This science recognises that natural selection shapes not only form and function but every aspect of an organism's life history. The vast diversity of organisms is not reflected only in their adult anatomy, but in also their entire life cycles.

Darwin, following Malthus, proposed that organisms over-reproduce and that natural selection weeds out the fittest for survival. The eminent Oxford ornithologist, David Lack, however, showed in 1947 that in birds clutch size has been adjusted by natural selection to maximise the number of fledglings that can be reared on the food available. Evolutionary ecology sees the enormous diversity of life histories observed in living organisms as the result of natural selection solving a series of 'decision' problems which face a developing organism. These problems include: at what age should it start to reproduce, how many times in it life should it reproduce, how much time should it allocate to reproduction as opposed to growth and maintenance, should its offspring be large and few or small and numerous but less likely to survive? Behavioural ecology looks at behavioural aspects of these problems, for example, allocation of time between foraging for food and breeding. In the Department, this approach is serving to unify the research of several colleagues who are looking at ecological and behavioural constraints on the life histories of birds and fishes in particular. Dr Neil Metcalfe, our one non-parasitological recruit to the Department since the mid-'70s, was appointed Lecturer to promote this endeavour. With concern for the environment becoming paramount, several colleagues are adapting their research interests accordingly.

In teaching, the loss of the medical class in 1988 caused some soul searching, this class for so long having constituted the *raison d'etre* for the Department, but it did help in alleviating teaching loads. I recall with wry amusement David Newth's complaint in 1977 that the staff student ration of 1:9 in his Department was unacceptable when the national average for biological departments was 1:7.8. Today the ratio is double what it was then. Though reduced in size,

Graham Kerr's Museum is still a notable feature of the Department and one of the few university Zoology museums surviving in the UK. Thanks to the enthusiasm of Dr David Houston and Ms Margaret Reilly, it has been extensively reorganised in recent years to include live animal displays (present occupants include chinchillas, poisonous frogs, a monitor lizard and a boa constrictor!). The Museum suffered a setback when shortly after our new invertebrate display had been opened by Sir David Attenborough, water from the hoses of firemen tackling the blaze destroying the roof laboratory above, poured through the ceiling and wrecked years of work. Luckily, the main exposed animal exhibits were saved from destruction, and the Museum was soon restored for use by students and public alike.

Retrospect: prospect

Since the foundation of the Faculty of Science one hundred years ago, remarkable changes have taken place in Zoology, in the focus of interest of the subject and in the ways in which teaching and research are practised. From a department of natural history consisting of a single professor, who taught all aspects of the subject, we have seen progression to a Zoology Department with a diverse staff of specialists revolving around an appointed professor, to a multiprofessoriate staff managed by a temporary head of department. The authoritarian Scottish Professor of Natural History has become a figure from the past.

Many of the dramatic changes in the conduct of research have taken place in my own lifetime as a professional zoologist. Individual research has effectively been replaced by team research, the single-author publication by multi-author papers. Simple techniques have given way to complex technologies involving use of equipment whose operation and maintenance requires a dedicated technologist. Computer-assisted collection of data can now take place upon a scale undreamed of 20 years ago, leading to an obsession with analysis and problem solving, a reduced propensity for synthesis and problem posing. Techniques are so time-consuming and change so rapidly that only those researchers unencumbered by administrative and teaching responsibilities can keep up with them. Bench research in universities is becoming the prerogative of the postdoctoral generation and one prominent biologist has predicted that by the end of our present century, the normal age of retirement for scientists will be 45!

From the past let us turn to the future. Why should the University retain zoology and other 'vertical' divisions of biology (botany, microbiology) when in many other universities biology is now split horizontally so that investigators who work at a particular level of organisation (molecular, cellular and so on) form a department? In my view there are several arguments in favour of retention.

First, there is a need to integrate investigations at different levels and zoologists, by operating at different levels but focusing on the whole animal, can do this. At the risk of appearing self-centred, I shall illustrate this point by reference to my own interests. The parasites I originally selected for special study were the trypanosomes, partly because the diseases they cause are so terrible and have a stranglehold on entire continents and partly because the future of the unique mammalian fauna of Africa (dear to all zoologists!) is intimately bound up with the ecology of the trypanosomiases. This led me to ask how these parasites can possibly survive the transition from one host (insect vector) to another host of a very different kind (mammal) and how they can possibly survive for so long in the mammalian host in the presence of an active immune response. The answers were, respectively, by cyclically activating and repressing the mitochondrion and by changing the antigenic nature of their surface. Further pursuit of these questions by molecular biologists has led from trypanosome mitochondrial activation to the discovery of RNA editing and from trypanosome antigenic variation to the demonstration of new mechanisms of gene expression. What started off as an ecological problem has ended up being studied at the other end of the biological spectrum.

This example illustrates other advantages of the zoological approach. By spreading their interests over more unusual animals, zoologists can discover new favourable material and new systems for horizontal biologists to work on; they can pose new problems. Where would developmental biology be without the sea urchin egg, genetics without *Drosophila*, neurophysiology without the squid giant axon, memory studies without the sea slug *Aplysia*? Zoology celebrates the staggering diversity of animal organisation and uses the exception to illuminate the rule. Recognition of this diversity is becoming extremely important, and I will explain why.

The heroic discoveries of the middle of the 20th century - the structural organisation of the cell, the pathways of cell metabolism, the genetic code, the mechanism of protein synthesis and so on, shifted the attention of biologists from the whole organism to the molecular level, and a fundamental unity of all organisms from bacteria to humans was revealed. The promise of all-pervasive understanding of life itself, induced the research councils to pour the contents of their depleted coffers into research concerned with description and analysis of molecular universals. Whole organisms and their diversity took a back seat as many universities combined departments which distinguished between animals, plants and microorganisms. One short-sighted devotee of the molecular approach even suggested in *New Scientist* that degrees in Botany and Zoology should be discontinued!

The situation is now changing as the molecular revolution moves on to consider the question of how, in spite of all their molecular similarities, have different organisms come to be so diverse. The diversity of organisms is back in the picture, indeed since the Rio Summit Meeting last year, 'biodiversity' has become a buzz-word! Biology is on the threshold of a transition comparable in magnitude to the molecular revolution of past decades.

One of the most heartening aspects of teaching zoology in a university in an age when the spirit of the universities has been reduced to a low ebb, is the popularity of our subject with students. This in large measure stems from real concern on their part for the future of the living world. The biggest challenge facing mankind in the next 50 years will be how to accommodate the inevitable massive explosion of the human population without destroying the rest of the biosphere. Conservation and the maintenance of biodiversity have to be an essential part of any strategy for sustainable development without degeneration of the environment. Zoologists hope to have a say in these matters.

Keith Vickerman F.R.S, is Regius Professor of Zoology at the University of Glasgow and Honorary Keeper of the Hunterian Zoological Collections.

Chemistry

George L. Buchanan and Samuel J. Thomson

Though the aims of the alchemists in their search for the philosophers' stone and the elixir of life were not fulfilled, their legacy was of fundamental importance to the craft of the chemist: it lay in apparatus, processes such as purification, crystallisation, calcination, the preparation of acids, bases, salts, metals and non-metals. Thus it was that the first Lecturer in Chemistry in Glasgow in 1747, William Cullen, taught a subject which was the hand-maiden of medicine and by 1892 the servant of the industrial arts: indeed this telling description of the latter is used in the 1891-92 *Calendar.*

One hundred years ago the teaching of chemistry, which had begun to assert its independence from medicine, was in the hands of the Regius Professor, John Ferguson, M.A., and his two Assistants, G. G. Henderson and J. S. Muir: they dealt with a dozen or so students. The Regius Chair was founded by George III in 1818, by which time chemical research was well established. However, during Ferguson's tenure of the Chair (1874-1915) the Department's interests changed direction. A contemporary has reported that, after he became Professor, Ferguson "never did any chemical research". He became interested in, and a recognised authority on, the early history of chemistry and he amassed a formidable collection of books in that field, particularly on Paracelsus.

After the University's move from the Old College in High Street to Gilmorehill, chemistry lectures were held in the easternmost lecture theatre of the Gilbert Scott building. Adjacent to it, on the site now occupied by the James Watt Engineering building, practical chemistry was accommodated in a high-roofed octagonal laboratory whose design was reputedly based on the monks' kitchen at Glastonbury Abbey. In 1891-92, lectures on systematic chemistry were delivered daily at 10am during the winter session. They commenced with the discussion of the general principles of chemical philosophy, the non-metallic and metallic elements and their compounds were 'treated of'. The latter part of the course was devoted to organic chemistry. There were tutorial classes twice a week and an analytical laboratory open daily during the winter and summer sessions from 10am to 4pm. In the laboratory, besides instruction in analytical chemistry, there was the application of chemistry to medicine and the industrial arts.

There was in addition a course on practical chemistry intended for medical students, given in the summer session. It treated the methods of testing bases and acids, the detection of poisons and the preparation of pharmaceutical compounds. Attendance on the course of chemistry was required for the Certificate in Engineering Science and for the degree of B.Sc. The examination in chemistry had five compulsory questions and these reveal the nature of the subject a hundred years ago. Students were asked to explain terms such as allotropy, nascent state, catalysis, describe the preparation and properties of urea, the manufacture of iron and steel and describe the relationship between atomic weights and the chemical properties of the elements. Chemistry was at that time descriptive and with little theoretical content: from this it was to grow rapidly in the next century, for advances were being made in the subject which gave insights and understanding which were to transform it from a craft into an intellectual and philosophical science. Major advances were to come in thermodynamics, kinetics and studies of atomic structure.

Changes were also occurring in the early part of the century in student numbers. Accordingly, 'temporary' brick out-buildings were added, and these were subsequently adapted to meet the 1919 influx of ex-servicemen. At the same time, the Ordinary class had swollen to over 400 and lectures were delivered in two sittings. By the late 1930s, these dilapidated buildings (now shored-up with a wooden buttress) still provided all the accommodation available for teaching and research in Chemistry. Nor were they well equipped. As far as the undergraduate was concerned, the entire Department seemed wedded to thrift and 'do-it-yourself'. Some communal laboratory hardware was supplied but, for example, there were no electric stirring motors. Students made do with Stirling hot air machines driven by bunsen burners. At that time students were required to buy their own glassware, which was inevitably of the cheapest (soda) glass, and to provide their own box of accurate weights. All of this was available from local suppliers whose agents hovered at the rear of the laboratories at the beginning of each year to take orders. At the end of the 1930s there were no sintered glass filters or any apparatus fitted with standard interchangeable ground-glass joints; yet A. R. (later Lord) Todd encountered both of these in widespread use at Frankfurt in 1929. Glasgow students of that vintage had to learn the delicate technique of cork-boring and under these conditions, vacuum-distillation was as much art as science.

Just as the accommodation changed in the first quarter of the century so also did the subject. The clearly defined branches of the subject were recognised by the appointment of lecturers in organic, petallurgical and physical chemistry. These subjects developed independently for the first half of the century but in their progress in the second half of the century they became intermingled in a most fruitful way.

We propose therefore to begin with the growth of organic, then physical and inorganic chemistry.

Organic chemistry

If Ferguson's interests lay in books, his successor can be recognised as a main stream experimental chemist. George G. Henderson F.R.S. ('G. G.') had been Ferguson's Assistant and later Freeland Professor in the West of Scotland Technical College before being appointed to the Regius Chair at the University of Glasgow in 1919. At the same time, Thomas S. Patterson, who had been a Lecturer in the Department since 1904, was appointed to the newly endowed Gardiner Chair of Chemistry. The two men had very different interests. Henderson was already a successful organic chemist with a reputation in terpene chemistry and he went on to enhance his and the Department's reputation in this field. On the other hand Patterson, like Ferguson, was really a historian with a passing interest in chemistry. His lectures to the Honours class were devoted to the history of chemistry and the history of optical activity. Nonetheless, the basic teaching of the subject was in the safe hands of Charles Buchanan and the charismatic S. Horwood Tucker. In addition, creditable research was being published by J. D. Loudon and T. S. Stevens (who was later to be elected to the Royal Society) and a steady trickle of students were sent to postgraduate study in Europe and the USA. One such, A. R. (later Lord) Todd, Nobel Laureate 1957, chose to move to Frankfurt for his Ph.D. work.

Despite their differences, Henderson and Patterson recognised the pressing need for new laboratories and collaborated in the planning of a new building for Chemistry at the western end of the campus. Work began in 1937 to a plan submitted by the architects, T. Harold Hughes and D. S. R. Waugh. After visiting several continental Chemistry departments they produced "the finest and best-equipped department in the UK at that time". To be exact, only two-thirds of the plan was completed by 1940 because of war-time restrictions, but it incorporated a number of novel features; a central store, a 400-seat lecture theatre with projection facilities, special isolated laboratories for dangerous experiments which involved risk of fire or flooding. In addition, the top floor accommodated the entire Chemistry collection of the University library.

The new Institute of Chemistry, as it was named, was a dramatic improvement on the old department with its broad corridors, high ceilings, excellent lighting and built-in services. Indeed, its parquet flooring and solid teak benches gave it an air of opulence.

In 1937, G. G. Henderson retired from the Regius Chair to his beloved Isle of Harris, but the University had found an able successor in Professor George Barger F.R.S., a distinguished natural product chemist. Sadly, Barger died in Switzerland just over a year later, before he could make his presence felt.

It was against this background that J. W. (later Sir James) Cook F.R.S. was appointed Regius Professor and Director of the Chemical Laboratories. He arrived in time to move into the new building in 1940; or at least into the two-thirds which had been completed before wartime restrictions halted building work. Cook's appointment in 1940 was a milestone in the Department's history. He had come from full-time research in a small London institute to take responsibility for a large Department with some 500 students - soon to rise to 1,000. It was a daunting prospect to lead one of the UK's largest departments but Cook had a clear orderly mind and he was able to view the situation with a fresh eye. He found a strangely lop-sided staff in which his only professorial colleague, T. S. Patterson, was likewise an organic chemist but whose main interest was in the history of chemistry. When Cook arrived Patterson had already established the syllabus and allocated the teaching duties. Cook selflessly undertook the 'chore' of teaching the Ordinary class, a large class with a reputation for rowdiness, whose lectures were mainly concerned with physical and inorganic chemistry.

When Patterson retired in 1942 Cook began to reorganise the teaching and re-equip the laboratories with modern apparatus. Most significantly, he appointed a crystallographer, J. Monteath Robertson, to the second (Gardiner) Chair. It was a move that was to bring rich rewards to the Department.

As to the teaching, the Chemistry Honours course which Cook inherited from Patterson was somewhat unusual. It included compulsory courses in 'the history of chemistry' and 'the history of optical activity', these being Patterson's research interests. Students were also required to pass simple written examinations in French and German. The final year was one of specialisation in which **one** of the three main branches of chemistry was chosen and studied in depth. The final Honours examination consisted of three papers in which all questions were compulsory. Before it became common practice, Glasgow students were also required to undertake a small research project during the final year, to submit a thesis on their work and to deliver a mini-lecture on the subject. Most unusual of all, Honours students in organic chemistry were required to **prepare** and **deliver** about 70% of the lectures in their final year. The lecture syllabus was divulged in good time, names went into a hat and the unlucky 'winner' had to deliver the lecture to the assembled class. A staff tutor supervised the proceedings and usually added supplementary lecture notes of his own. As a teaching device this practice had some merit but it had obvious demerits and Cook abandoned it. He likewise removed specialisation and brought in a broadly based Honours course. In addition Cook proved to be a splendid administrator, a talent that eventually lured him to two vice-chancellorships. He persuaded Dr James Bell to be the first Assistant Director of Laboratories.

Other changes were desirable, to foster research, to bring in young staff, particularly in physical chemistry, and to introduce new instrumentation. Though in 1942 Britain was fighting for its life, he and Robertson began to redress the staff imbalance by making appointments designed to broaden the Department's expertise in theoretical (T. H. Goodwin), physical (J. C. Speakman), physical-organic (J. D. C. Brand) and inorganic (D. S. Payne) chemistry. The renaissance of the Department dates from this time.

Cook's reputation had been made in the Royal Cancer Hospital in London by his success in isolating and identifying the carcinogenic constituents of coal-tar. When he moved to Glasgow he continued his work on carcinogenic hydrocarbons with G. M. Badger and Miss R. Schoental. He also initiated a search (with W. Carruthers) for natural carcinogens in petroleum. However, he was also turning his attention to the feasibility of inhibiting cancer, and in particular to the report that colchicine, the alkaloid from the Autumn Crocus (colchicum autumuale L.) had produced remissions in experimental tumours.

But colchicine's structure was an enigma which had long defied solution. In tackling this problem (with J. D. Loudon) Cook pushed Glasgow into the forefront of organic chemistry for its eventual solution, in which Glasgow played a large part, led to the discovery of a new and fundamental molecule, tropolone. In this spin-off, the Glasgow department continued to play a key role through the work of Professor J. M. Robertson and R. A. Raphael.

Meanwhile, work was progressing on building the remaining wing of the Institute of Chemistry - but progressing slowly for, in preparing the foundations, the contractors uncovered old, uncharted mine-workings said to harbour a ghost. The work was finally completed in 1954 and, in the same year, Professor Cook departed to become Vice-Chancellor of the University of Exeter.

During his years in Glasgow, the Department welcomed two distinguished refugee scientists; Vladimir Vand and Erich Clar. Even then Clar was a recognised authority on polycyclic aromatic hydrocarbons and he continued his work in Glasgow. He was awarded the Kekulé medal of the East German Chemical Society in 1965, the Kekulé Centenary year. Dr Vand contributed both to electron microscopy and computing.

New appointments were made also in agricultural chemistry (F. Moon) and in electrochemistry (G. Nancollas) whilst in organic chemistry G. Eglinton took charge of the first infra-red spectrometer - the beginning of the 'spectroscopy revolution' which was to change the face of the subject.

Losing a man of Cook's calibre could have been a crushing blow to organic chemistry at Glasgow. But in appointing Professor D. H. R. Barton F.R.S. in his stead, Glasgow had found another 'giant' - arguably the greatest organic chemist of his generation. In later years he was knighted and in 1969 he became a Nobel Laureate; but in the three years he spent at Gilmorehill (1954-57) he galvanised the organic department. It became a hive of activity and in the research laboratories the lights burned well into the evenings. Post-doctoral collaborators came to Glasgow from all over the world and even senior American scientists began to 'drop in' via Prestwick, on their way to London. Glasgow became an exciting place in which to work and the source of new ideas.

In Professor Robertson, Barton found an ideal colleague, whose research work was complementary to his own. Robertson had gained access to his first computer and for the first time this enabled his crystallography research group to solve complex chemical structure problems rapidly. Barton too was involved in determining molecular structure by organic techniques and the two groups made a powerful team. Robertson had begun his research career as an organic chemist, under G. G. Henderson, which may explain his life-long interest in natural products. Back in 1926 he had published some (inconclusive) experiments on the chemistry of caryophyllene, a terpene from clove oil. Ironically, it was Barton who finally unravelled this complex puzzle, some 30 years later; and Robertson who returned to this topic as a crystallographer in 1955, confirmed Barton's deductions.

When he came north, Barton brought some senior colleagues with him and some of these, notably K. H. Overton and C. J. W. Brooks elected to stay and added their own talents to the Glasgow staff. But all too soon Barton was lured back to Imperial College in London. With his departure, in 1957, Professor R. A. Raphael from Belfast was appointed to the Regius Chair. He had previously been a member of the teaching staff (1949-54) all of whom knew him well, and were equally well-known to him. It made for a smooth transition and the organic department enjoyed a 'family' atmosphere. He was a good listener, and much of the departmental planning was based on consensus, following wide consultation. At that time, finance was available from a benign government, and he took the opportunity to acquire the latest state of the art equipment. At the same time too, the teaching programme was upgraded to give senior undergraduates 'hands-on' experience of these latest techniques - mass spectrometry, nuclear magnetic resonance spectroscopy and gas chromatography. It is perhaps no coincidence that this period saw a rise in student numbers. The Ordinary class was too large for the largest lecture theatre and had to be divided. The overflow group was taught in another theatre either by a separate lecturer, or via a TV relay from the main theatre. At this time too the Honours Class swelled into the 1960s and '70s with a peak of 74 in 1970. Interestingly, quality did not suffer for the class of 1970 yielded 22 first class and 45 second class degree passes.

Raphael's research interest lay in organic synthesis and his swelling band of post-graduate students recorded many notable successes. Under the Raphael-Robertson regime Glasgow continued to be a recognised centre of excellence. Perhaps its greatest success at that time was continuing to turn-out high calibre chemistry graduates and post-graduates. Even yet, Glasgow crystallographers and organic chemists are to be found in senior positions throughout North America, Europe and Australia. During these eventful years, 1962 stands out as the Department's 'annus mirabilis'. In that year Dr. D. W. J. Cruickshank was appointed as its first Joseph Black Professor, Professor Raphael was elected a Fellow of the Royal Society, Professor Robertson was elected President of The Chemical Society, the new organic research floor was opened, Dr. A. I. Scott (one of our graduates) resigned his Lectureship to become Professor at the University of British Columbia (Vancouver) and J. M. L. (Jimmy) Cameron, our renowned micro-analyst, was capped B.Sc. (Honoris Causa) to everyone's delight.

But Raphael's success in Glasgow had been attracting attention elsewhere and in 1972 it resulted in his appointment to Cambridge.

Physical chemistry

The first incumbent of the Lectureship in Physical Chemistry in 1904 was Frederic Soddy (1877-1956). He had worked for two years in McGill University with Ernest Rutherford on the explanation of radioactivity showing that the disintegration of one substance led to the production of another with the release of a particulate radiation. [Th → ThX → Thorium Emanation]. In 1903 with Ramsay in London he demonstrated that the particulate radiation, when captured, consisted of helium. From this background he came to Glasgow in 1904 at a time when there was by no means universal acceptance of the basis or validity of the Periodic Table of the Elements.

The Dutch lawyer Antonius Johannes van den Broek had in 1912 suggested atomic number as the basis for arranging atoms in the Periodic Table. This concept was however thrown into disarray by the great variety of naturally occurring radioactive elements which were soon to be discovered and which could find no reasonable place in the Periodic Table.

In a demanding series of experiments between 1904 and 1914 in Glasgow, Soddy clarified the relationships of the plethora of natural radioelements with the Periodic Table. Many of these were inseparable chemically (eg, mesothorium I, radium and thorium X) and Soddy took the bold and imaginative step of declaring them to be chemical identities, recognising that elements of different atomic weight may possess identical properties. For these Soddy coined the word isotopes. He proceeded from there to his alpha-ray rule (1911) in which expulsion of an α-particle results in a lighter element in a next but one position in the Periodic Table. In 1912 he assigned Alexander Fleck to the task of understanding β-emitters. The one step changes of position in the Periodic Table led Soddy to pronounce that successive places in the Periodic Table correspond to unit differences in the 'intra-atomic' charge. He thereby established once and for all the basic concept of atomic number as fixing positions for atoms in the table. These fundamental discoveries together with later work on branching of radioactive series led to the 1921 Nobel Prize.

Thus when chemistry was given its basic concept of the nuclear atom surrounded by electrons interpretation became possible for atomic behaviour, ionic compounds, and metals.

Soddy's validation of the Periodic Table based on the atomic number of nuclear atoms laid the basis for much of the teaching in physical and inorganic chemistry in the 1920s and 1930s. This centred round ionic models for salts and salt solutions. Metals were visualised in terms of positive ions in a sea of electrons. Research in the Department in physical chemistry in this period involved deuterium isotope studies by J. Bell, dielectrics by G. Thomson and optical rotary dispersion by S. T. R. Mitchell.

The 1939-45 period was one in which wartime controls overshadowed everything. Evening working was not possible because of the blackout; staff and students were organised into teams of firewatchers, who guarded the buildings each night. The Department experimented briefly with a telescoped Honours course by reducing one of the summer vacations to two weeks. However, it was an exhausting experience all round and it was abandoned. The Henderson laboratory was commandeered by the Admiralty for steel analysis. This work required large amounts of concentrated acid and it was no surprise when, at a later date, the under-floor electric conduits were found to be badly corroded. Eventually, by 1945, controls were eased and it was possible to resume and recruit more staff. At the same time the teaching load increased for in addition to the long standing medical chemistry class, arrangements were made to teach chemistry to arts, agriculture, dental, engineering and veterinary students.

In spite of these war-time difficulties Professor J. M. Robertson was able to found his school of X-ray crystallography though he had to await the post-war years before he could develop his strategy for expansion in breadth of both physical and inorganic chemistry. Robertson's Ph.D. studies in organic chemistry had great significance for his future. They revealed problems in elucidating structures of complex organic molecules, particularly among natural products, where chemical means were inadequate. He realised that structures deduced from chemical experiments, however elegant, revealed only plausible arrangements of atoms. What he was to invent was a way in which to 'see' atoms and to measure exactly the architecture of organic molecules.

Robertson was aware that Sir William Bragg and his group at the Royal Institution were using X-rays in diffraction studies of the structure of inorganic salts, silicates and metals. He joined this group in 1926 as a Carnegie Fellow to begin his life's work in structure determination of organic crystals. The problems he faced were formidable; there were no simple transfers of existing diffraction methods to structural studies of organic crystals. By a combination of diffraction, chemical intuition and laborious calculations he succeeded in producing structures for some twenty compounds in the 1930s. These included naphthalene, anthracene, durene, hexamethyl benzene, dibenzyl, stilbene, tolane, resorcinol, benzoquinone and oxalic acid. The dimensional results were of great importance in development and refinement of theoretical chemistry and they represent Robertson's first great scientific achievement.

In Robertson's mind there were still fundamental problems in structure determination by X-ray scattering and the task that he then undertook was daunting.

He was aware early in the 1930s that diffraction methods applied to organic solids were confirming structures already known and not necessarily producing structures which were unique and unambiguous. Robertson's solutions to this problem were landmarks in the development of X-ray crystallography, forming the basis for a host of subsequent studies. X-ray scattering produces beams whose intensities and hence amplitudes can be measured. This information cannot be translated into atomic structures without knowledge of both the amplitudes of the scattered beams and their experimentally inaccessible phase constants. It was the phase problem which Robertson mastered in three profound papers in the *Journal of the Chemical Society* in 1935, 1195: 1937, 219: 1940, 36, the latter two with Ida Woodward. In these papers Robertson developed isomorphous replacement in the phthalocyanine series to compare molecules with H, Ni and Cu at their centres. This allowed him to find phase constants and absolute structures. When Pt was placed centrally in phthalcyanine the heavy atom method gave known phase constants dependent only on the heavy atom. The positions of all other atoms could then be determined. He says the very assumption that atoms were present in the molecule was unnecessary.

Robertson's 1939 paper in *Nature* (Volume 143, 75) revealed his vision for the future. In it he discusses the limitation of the Patterson method and predicts that even the insulin structure might be solved by the heavy atom method if its few zinc atoms were to be replaced by mercury: this prophecy was fulfilled 30 years later.

Robertson was appointed Gardiner Professor of Chemistry in the University of Glasgow in 1942. This led to a third great contribution to structure determination for in Glasgow he established a school of crystallography which was to become world famous.

The molecules previously chosen by Robertson in his studies were essentially planar for the computational methods available were restricted to two-dimensional studies. This was to change in the late 1950s. Such was Robertson's foresight that the first computer, DEUCE, which came to the University of Glasgow was housed in Chemistry. This opened the way for three-dimensional studies and fulfilled Robertson's wish to study the structures of natural products and many other classes of molecules. Thus the school Robertson founded in Glasgow in the mid-1940s was poised to make a quantum leap in the variety of problems which could now be addressed. About a hundred papers appeared in the next ten years: this was a remarkable contribution to science made in the final years of his professional life before he retired in 1970.

At the close of this decade Professor Robertson retired from the Gardiner Chair at the end of an outstanding career, and one of his pupils Professor G. A.

Sim returned from the University of Sussex to lead the Group. Happily, Professor Robertson returned to the Department in 1980 to participate in a special conference held to celebrate his 80th birthday and attended by a large gathering of his former students and collaborators, now pursuing successful careers in Europe and North America.

The Computing Department's English Electric DEUCE computer was housed in Chemistry. Just how important it was for the structural chemists is revealed in the October 1959 research report of the Chemistry Department. "The four month period preceding this report has been one of the most remarkable in the history of the Department. Never before has a like period seen such a massive fructification of discovery of crystal structures."

Chemistry's use of the computer at that time exceeded all others for, in a typical period of the 430 hours running time for research available in the three months to March 1959, 300 were used by 16 members of the Chemistry Department. Such was the extent of computing in crystallography that by 1960 it was proposed that there be a spring-clean of the Hollerith cards, evidently knee deep by this time, for the estimated number in Room 262 was over 100,000.

X-ray structural studies were complemented by electron microscopy under I. M. Dawson. Images then were made from shadow-cast specimens of crystals, metal particles, carbon, bacteria, ribonuclease and lysozyme. A fundamental observation made was that of spiral growth of crystals and by 1959 Moiré magnification had revealed the hexagonal layer nets proposed by Bernal in 1924.

The polymer group was instituted by N. Grassie with major interests in degradation. The 1950-60 period covered radical reactivities, lifetime of radicals, weak links and degradations induced by ozonolysis, heat, oxidation and photolysis.

J. Bell and R. I. Reed introduced one of the first commercially available mass-spectrometers for isotope analysis in 1949. R. I. Reed went on to develop its application to determination of bond energies in organic molecules by electron impact. The studies were extended to dissociation of organic molecules with the aim of structure elucidation.

Radiochemical studies of surfaces and solids began in 1957 under S. J. Thomson soon to be joined by K. C. Campbell, G. Webb and I. M. Hoodless. They exploited the ability of radiation-detection methods to 'see' and follow the tiny numbers of atoms and molecules adsorbed on surfaces during catalysis, diffusion and trapping in solids. Before such studies on surfaces were made atomic and molecular populations had only been estimated by informed guesswork. J. N. Sherwood's contribution to the group was to accomplish the growth of single crystals of anthracene and make the first determination of a diffusion coefficient in an organic crystal. In 1963 J. A. Cairns studied focused collisions in solids again applying radioactivity detection.

Theoretical chemistry was to begin with T. H. Goodwin's studies of bond lengths and orders in cyclic hydrocarbons. By 1963 there were studies in the Department on electronic structures of molecules and solids, semi-empirical calculations of inter-molecular potentials, d-orbitals and molecular orbitals.

Kinetics was the field of study instituted by E. Gelles. Underlying the work of the group was his interest in the role of metal ions in biological reactions. Model systems were used in studies of decarboxylation and hydrolysis in an attempt to relate these to enzymatic reactions. Topics appear in the group research under amine-catalysed decarboxylations, acid-catalysed solvolysis, metal ion-catalysis, and isotope effects in decarboxylation.

Kinetic studies formed a part of the electrochemistry group directed by G. H. Nancollas which investigated crystallisation and dissolution of sparingly soluble salts, thermodynamics of ion association. R. Paterson joined the group to begin work of the physical chemistry of inorganic ion-exchange materials and Helen S. Dunsmore began work on uranium salts and the programming of electrochemical calculations on DEUCE.

In 1960 spectroscopy entered a new age which was to have repercussions over all of Chemistry when A. L. Porte installed a nuclear magnetic resonance spectrometer. This was soon followed by instrumentation for electron spin resonance and nuclear quadrupole resonance. J. K. Tyler constructed a Stark-modulated microwave apparatus for dynamical, dimensional and structural analyses of small gaseous molecules. D. W. J. Cruickshank appointed for theoretical studies to the Joseph Black Chair introduced electron diffraction observations of small molecules. Infra-red spectroscopy already used as an analytical tool became a research interest in its own right under L. Barron who began his research on laser-Raman and polarised infra-red spectroscopy. The 1960s also saw the arrival of J. H. Dymond who came to study the thermodynamic properties of fluids, M. S. Child, G. Doggett, R. Emanuel and B. Webster to pursue theoretical studies.

Inorganic chemistry

Inorganic chemistry in the 1930s and 1940s was dealt with in teaching as a descriptive subject in lectures and as an analytical subject in laboratories. In research Gibson pursued studies of sulphur and selenium chemistry as well as perfecting elegant methods and apparatus in microanalytical chemistry. Mair's research was into the formation of heteropoly acids.

Three important texts were to signal a change in direction in inorganic teaching: they were A. F. Well's book on structural inorganic chemistry of solids, H. J. Emeléus and J. S. Anderson on modern inorganic chemistry and L. Pauling's text on chemical bonding and valency.

The resurgence of interest in inorganic chemistry can also be traced to the wartime research aimed at producing nuclear weapons. Inorganic chemistry played an essential part in this project equal to that of physics, metallurgy, engineering and electronics. The chemistry of uranium, plutonium and the fission products became of vital importance and its peaceful legacy was the birth of a new inorganic chemistry which now had at its disposal a range of new radioactive elements. Neutron activation extended it further.

The opening of these new fields brought D. S. Payne to Glasgow in 1949 to add another strand to the existing analytical and preparative chemistry. Studies began in phosphorus chemistry with conductance and dissociation measurements of PF_3Cl_2 and pyrolysis of phosphites using a thermal balance. Infra-red and mass spectrometry added to the physical methods of analysis available to the inorganic chemist. G. S. Harris joined the group to study perfluoroalkyl derivatives of phosphorous and organo-arsenic halides.

These extracts from the research reports in the Department illustrate the post-war growth. There was to be an even greater impetus given to the subject by the appointment of J. M. Winfield and D. W. A. Sharp (Ramsay Chair).

Under Sharp's direction in the 1960-70s, inorganic chemistry grew as a separate and substantial subject. His interests lay in the chemistry of phosphorus-fluorides and their derivatives, their preparations and structures: the latter by nuclear magnetic resonance, infra-red and Raman specroscopies. A major interest was also in sulphur-fluorides some of which had potential as inert liquids and gases, others as fluorinating agents. A third area of research was in complexes of transition-metal salts with heterocyclic bases.

J. M. Winfield's work was on the synthetic and spectroscopic studies of transition metal fluorides and their complexes. High oxidation state species such as $WF_{6-n}(OMe)_n$ functioned as fluorination agents. Metal pentafluorides and their complexes with N, O and S ligands were also studied. A promising field which led to major successes was in reactions catalysed by metal fluorides or

Lewis acid fluorides in respectively perfluoro-olefin reactions and Friedel-Crafts reactions.

Kinetics and reactions of metal complexes in solution was the chosen field of A. McAuley. P. S. Braterman investigated metal-metal interactions in polynuclear metal carbonyls with development of the 18 electron rule into an 86 electron rule. This work was accompanied by vibrational and electronic spectral studies of metal carbonyls. Organometallic chemistry was represented by R. J. Cross's research on Ni, Pd and Pt. R. Keat's research was on phosphorus compounds of the type P-N-P and P-N-Si. A. Lane's interest was in solid-solid vibrational spectroscopy.

A school of geochemistry was founded by A. Walton based on natural and artificial radionuclides: when he left he was replaced by M. S. Baxter.

Housed in inorganic chemistry yet stretching across all of chemistry was A. H. Johnstone's Chemical Education Unit. They looked searchingly, critically and constructively on the whole process of learning chemistry and the way in which it is examined. They ranged widely over the alternative syllabus in schools, University syllabus content and its order of presentation, use of language in learning, the value of project work and fixed response questions in first year examining.

Agricultural chemistry

This sub-department of Chemistry was founded after the War: its permanent staff, usually about four in number, teach a separate Honours degree. Together with a dozen or so postgraduate and postdoctoral research workers the group researches widely in agricultural, food and environmental chemistry. It is a group widely consulted by outside bodies and is the target for many overseas visitors. Though not its main purpose in life, it has nonetheless maintained Chemistry's tradition of analytical chemistry. Its methods bear little resemblance to those of the past for atomic absorption, neutron activation chromatography and the like have superseded the early wet methods of analysis.

The 1978 research report shows the group leader as Dr. H. J. Duncan. The topics then were as follows. Land reclamation had been studied for five years in the form of work on the greening of the pit bings in Scotland. The hostility of these bings towards plant growth had been ascribed to the presence of iron sulphides and their oxidation products and the search was on to find methods of inhibiting the oxidation. This was coupledwith intensive studies of other trace metals. Recent studies had also been made of oil and tar-contaminated land.

The behaviour and fate of herbicides was another major field of study. This was a research which involved biochemistry, metabolism, methods of application of herbicides, their erratic behaviour, their fate in the food chain and the soil.

Similarly connected with food was their work on potato sprout suppressants. Here the problem was to study sprouting in stored potatoes and its inhibition. More importantly the study continued into examining residual quantities of suppressants in food products.

Forestry occupied the group in their study of the failure of spruce trees to grow in what seemed to be promising soils derived from basalts. The problem was studied and identified as being caused by release of toxic amounts of aluminium which appeared to inhibit uptake of phosphate.

The origins of life

The Department is not solely concerned with the practical. Dr. Cairns Smith's interest is in the origin of life and he has published two books on the subject. He has been critical of current hypotheses which are based on the production of DNA precursors in electrical discharge experiments (Miller) and has sought alternative pathways.

Miscellaneous

Our account of the three major branches of chemistry together with agricultural chemistry would be incomplete without a commentary on the other aspects of Chemistry: how was it governed, what of its students and technicians, what gave it life and purpose?

The management of so large a Department from the 1960s onward consisted of the Director and the two principal professors who met from time to time with the Assistant Director, Dr James Bell. But by 1972 when Dr Bell was succeeded by Dr S. J. Thomson, the staff of two secretaries who had been sufficient to deal with Cook's requirements in 1940 had expanded to provide each professor with two secretaries. In addition Miss J. Ann Boyd, one of Cook's original secretaries was now an Administrative Assistant with overall responsibility for all secretaries and became a key figure in the Department. She played an increasing role in preparing Department documents and returns and held sway over the accurate preparation of examination papers and records.

When Professor Raphael, who had been Director from 1970 to 1972, was enticed to Cambridge he was succeeded as Regius Professor by G. W. Kirby. Professor D. W. A. Sharp became Director in 1972 and served until 1978 when he declined a further term. In an unprecedented move, the senior members of the non-professorial staff met and elected S. J. Thomson, now a Titular Professor, as the next Director of Laboratories. The University Court gave its blessing to this initiative and accepted the suggestion that Dr. G. L. Buchanan should be the Assistant Director. The Chemistry Department now boasted three appointed professors, five titular professors, five readers, 16 senior lecturers and 25 lecturers. Dr. Buchanan's successor as the last Assistant Director was Dr. A. J. Baker. Professor Thomson was succeeded in 1987 by Dr. R. V. Emanuel following a ballot, and by the same process Professor G. Webb became Head of Department in 1990.

Today, the views of our students come to us from tutorials and laboratory sessions, and in a more formal way through elected representatives of the various years.

The Department owes a debt to its technicians without whom it simply would not have functioned. We illustrate our point by choosing those who made exceptional contributions.

The first was Alexander Fleck who assisted Soddy in the later stages of his work. Fleck was the son of an Ayrshire coal merchant who began as a 'lab boy' in Natural Philosophy but moved to Chemistry at Soddy's instigation and rose, via evening classes, eventually to a Lectureship in Chemistry and collaborated with Soddy in elucidating the β-decay phenomenon. He later showed the same flair in industry, becoming Chairman of ICI and Baron Fleck of Saltcoats.

In the 1930s the lab boy was a feature of the Department. He was a 15-16 year old trainee, who skulked in a corner of the lab, ostensibly to do odd-jobs and run errands, but most of them were masters at doing as little as possible and few stayed long enough to acquire any skills. They are now a vanished race. However, one such (James M. L. Cameron) was trained by Dr. S. H. Tucker to carry out analyses and became an internationally acclaimed micro-analyst. Another who became Chief Technician was James Findlay. It was customary for the lectures to the Ordinary class to be illustrated by lecture-bench experiments, introduced in the main for sound educational reasons, but on occasion as a respite from note-taking. Thus they varied from the simple illustration to the spectacular (and dangerous) ether-gutter fire, which always drew generous applause. But any lecturer who embarked on a bench experiment without first consulting Findlay was risking disaster. Jimmy Findlay had a note book, full of stratagems to ensure that the experiment 'worked'. But he did much more, he supervised a training course for the junior technical staff.

Chemistry was fortunate in its Library which was a branch of the main Library. It was served by a succession of University librarians whose knowledge of the contents of the Library was astonishing and whose patience and helpfulness was remarkable. A reading room, designed by Ian Kay, was added in 1982: it was paid for by a bequest from the estate of George Lean for which the Department was most thankful.

A notable feature of the Department of Chemistry over the past 70 years has been the existence of the Alchemists Club, a flourishing student chemical society and one of the oldest in the UK. It has coexisted on a friendly basis with the staff of the Department but is run by the students in a spirit of fierce independence. We are told that in the 1920s, alchemists' meetings were held outwith the University in the evenings, and AGMs were evening-dress affairs. Today, dressing is less formal but the President still takes the chair wearing his presidential gown.

Over the years, its activities have consisted of regular meetings at which invited speakers addressed the Club on a variety of subjects, serious, trivial, scientific or simply entertaining. These were integrated into the Department's own programme of weekly lectures from a wide range of distinguished visitors.

The affairs of the Alchemists' Club are managed by the Council, membership of which is in itself an education in political craft. In the 1950s the Alchemists' Club obtained Union permission to run Friday night dances in the Students' Union; it was a money spinner. The Club made money, and for a time it was able to underwrite the annual Science Faculty ball. But the Medico-Chirurgical Society challenged their monopoly and a struggle ensued for control of the Union Board; the Alchemists organised the sister societies in the Science Faculty into a 'Council of Scientific Societies' to out-vote them. In the event neither won, for the Union decided to take over the Friday dances.

The Department of 1980, in contrast to 1892, was one of the largest and most dynamic in the UK. Staff had reached 50 members with 150 postgraduate and post-doctoral workers engaged in a wide field of research. It was a well-found Department with modern equipment covering 26 areas of research, an excellent library and extensive supporting technical and secretarial services.

A feature of recent times has been the scarcity of research funds. Although we have access to the Loudon Bequest Fund - a legacy from the late Professor James D. Loudon for research in organic chemistry - it does not go far. The speculative research of the 1970s has given way in the 1990s to research which will attract sponsorship. In 1980, for example, 23 external bodies gave financial support.

A glance at the research activities of the Department in 1978 gives an indication of the direction(s) in which departmental interests have developed in modern times.

Aside from the long-standing involvement with the chemistry of plant products, there is now a significant leaning towards biological topics; the investigation of Indian medicinal preparations; the study of enzyme mechanisms and their detailed stereochemistry, using isotopic labelling techniques; the chemical modification of antibiotics such as penicillin and cephalosporin with a view to improving their clinical properties.

In addition, the Department has always been renowned in the field of chemical synthesis and attempts are being made to synthesise biologically important substances which occur naturally in minute amounts. Allied to this is the question of how the plant itself goes about producing complex substances, ie, their biosyntheses, and much work is directed at unravelling nature's methods. In a fundamental approach to organic synthesis, some work has been done on devising new or more efficient synthetic methods.

In their research the groups which were founded in the post-war years continued to flourish but interests were now developing in biophysical chemistry. For example, the crystallographers were examining protein structures and conformational structure as well as continuing their work on organo-metallic compounds, theoretical studies of X-ray diffraction amidst a host of other topics. The electrochemists were examining electron and proton carriers in lipid membranes. D. J. White had opened up the study of organic molecules using molecular mechanics and molecular graphics with particular interests in polypeptides. Biophysical chemistry under A. Cooper was concerned with microcalorimetry and biological systems: the visual process and protein folding were the prime subjects. This change in direction was further emphasised in 1990 when Professor Sim retired from the Gardiner Chair to be replaced by the protein-crystallographer, Professor Neil Isaacs. The appointment of C. J. Gilmore brought to the Department leading edge research on the maximum entropy statistical approach to X-ray and electron scattering methods.

1980 - 1992

More recently, the management of the Department has become increasingly difficult as financial constraints have begun to dominate all planning and 'accountability' exercises have consumed more and more staff time. In addition, Dr Emanuel and Professor Webb have lately been faced with the task of completely refurbishing the Department's services as well as surrendering some 30% of its space to other users - while still maintaining the Department's full programme of teaching and research. It has been a noisy, trying, troublesome experience, but a modernised Department is emerging.

Chemistry, like all other departments, was deflected in part from its true purpose. A new vocabulary appeared - unit costs, staff-student ratios, area norms, funding crises, returns, projections, early retirals and the like - with accompanying efforts to attempt to measure efficiency. No doubt the present Head of Department is now wondering how to accommodate his 650 first year students.

In spite of the change of climate towards accountancy and monetary targets the Department held firm to its drive both to improve its teaching and to pursue its research. Visit it today and one is struck by a rebirth in Chemistry in the shape of a fine new Department created by George Horspool with the shell of the original (listed) building. Talk also to the staff and the same interest and dedication is at once apparent.

The Phoenix is arising from the ashes after Cardinal Newman's holiday.

George L. Buchanan is a Reader, and Samuel J. Thomson an Emeritus Professor, in the Department of Chemistry at the University of Glasgow.

Geology

Bernard E. Leake with archival research by Graham P. Durant

The exact beginnings of teaching and research in geology in the University of Glasgow are lost in the mists of time but it is known that the Hunterian Museum, with its magnificent collections of minerals, rocks and fossils, which was left to the University by Dr William Hunter in 1783, arrived in Glasgow in 1807. George III had been a friend of William Hunter and royal recognition of the value of the Museum and of its lack of a keeper seems to have prompted the founding, in 1807, of a post to fulfil both the needs of the Museum and of the University for teaching in natural history although the nature of this dual responsibility was however disputed bitterly by some members of the University and the first appointee despite the apparently clear "intention of the Crown to conjoin post of Professor of Natural History and Keeper of Museum or Repository of Natural Curiosities in the University".

The first appointment to the Regius Chair of Natural History in 1807 took place against the University's wishes and was of Lockhart Muirhead, a man of polymathic abilities who undertook the study of geology, including earth, water and air, and zoology - almost an environmental scientist. He was, however, also University Librarian, Lecturer in French and Italian, Registrar of medical students, briefly Keeper of the Hunterian Collections (this duty he objected to), almost Professor of Mathematics and appointed for 'all the days of his life'. But neither Muirhead nor his Assistant and successor (in 1829), William Couper M.D., made any visible mark on the developing science of geology nor did they put Glasgow on the geological map with distinguished and well-trained students of geology although with over 1,000 students (1,233 in 1826-7) the University was already large and popular. Couper was mainly interested in zoology and mineralogy in which he had valuable personal collections, and with his son James, was Joint Keeper of the Hunterian Museum, but his teaching seems to have been dull and uninspired so much so that Professor Thomas Thomson of Chemistry undertook 'without fee' to lecture in mineralogy and geology, "to make the students acquainted with what they ought to know", and it was Professor J. P. Nichol of Astronomy who delivered "brilliant lectures on cosmogony and speculative geology".

The first truly professional geologist in the University was Couper's successor to the Regius Chair, Henry Darwin Rogers, appointed in 1857. His middle name was after Erasmus Darwin, Charles Darwin's grandfather. Rogers (1808-66) born in Philadelphia, was an American and brother to the renowned W .B. Rogers, founder of Massachussetts Institute of Technology and President from 1862-70. H. D. Rogers had held Chairs in Chemistry and Geology in Pennsylvania but his geological survey of New Jersey, completed as state geologist, led to his appointment as Director of the Geological Survey of Pennsylvania. He devoted years of work, together with corps of assistants, to the survey of the State at the same time as his brother was mapping the State of Virginia. Both areas are dominated by the great Appalachian mountain chain and their joint publication of 1842 on the origin of these mountains was a "remarkable memoir", thought in 1866 "to be probably the most important of its class ever produced in America" and was followed in 1858 by H. D. Rogers' final report and enduring monument, the lavishly illustrated *Geology of Pennsylvania,* published in Edinburgh. The Rogers brothers conceived the antiforms and synforms which parallel the length of the Appalachians as the result of waves produced within the liquid interior of the Earth which flexed the overlying crust and then preserved the folds by intrusion of magma into them. Rogers published widely, studied earthquakes, but opposed Agassiz's glacial concepts. Rogers was an eloquent speaker, clear and graceful, able to explain difficult concepts and noted for the superb illustrations which accompanied his lectures and writings. In Glasgow as Professor of Natural History, which belonged to the Medical Faculty until 1893, he had to teach the medical students a three month course (in May, June and July) on anatomy and physiology. His initial, almost complete, ignorance of formal zoology had to be hastily removed by a week's concentrated cramming from T. H. Huxley together with the loan of a set of lecture notes! In February, March and April he taught the geology students and the remaining six months were available for travel during which he several times visited America. It is clear that geology was treated as a separate subject within Natural History by Rogers. Field excursions were included in which Rogers' command of language and extensive field experience in Palaeozoic rocks ensured excellent instruction but he himself never seems to have become interested in research in British rocks nor involved himself in the nascent Glasgow Geological Society founded in 1858.

When Rogers died in 1866 he was succeeded by another geologist with survey experience, "the remarkable John Young, untidy, impetuous, scornful, flamboyant, brusque, outspoken, bearded and an exhibitionist in his dress and enormous slouch hat". He graduated M.D. in the Edinburgh University Medical Faculty in 1857, practised for four years and then in 1861 resigned to enter the Geological Survey of Scotland without any formal geological training. After five years mapping in the lowlands and southern uplands of Scotland, he left the Survey, partly because of a painful fractured knee-cap. With his training in the Survey and his extensive knowledge of rocks and fossils gained under the tutelage of the leading geologists of the day (A. & J. Geikie, A. Ramsay and Ben Peach) he held great promise of promoting geology as a major science in Glasgow and did indeed introduce a graduating course in 1872 when the B.Sc. degree was established. However his geological enthusiasm faded, he became immersed in medical zoology and various University activities and never achieved any substantial piece of geological research. This was probably because of his wide-ranging, unfocused interests. He organised practical work in zoology, he was an authority on old books and coins, he sketched and painted, and he studied foreign literature; indeed there was hardly a branch of science or letters in which he was not widely read. His published geological work was mainly in palaeontology and glacial geology; in particular he made a large collection of glacial erratics discovered during the construction of the University building in Gilmorehill in 1869-70.

Brilliant, eccentric, full of sardonic wit, Professor John Young (1835-1902) was not a good teacher and was in retrospect called the "bad John Young". His lectures in geology were unconventional, never dull but not designed to instruct in a systematic way. He considered "lectures were a great waste of time and reading by the students would have been better". He gave the impression of hurrying into the classroom and impatiently addressing himself to an apparently hateful task - no-one set a lower value on his lectures than he did himself: "he must perforce, God help him, deliver them and the students must, the pity of it, endure them". Increasingly he withdrew from geology teaching and passed it onto his Assistants, one of whom fortunately was 'the good John Young'. However the attachment of the Honeyman-Gillespie bequest to the Chair of Natural History in 1876 increased his salary from £440 to £650 a year but imposed the condition of giving not less than 52 lectures a year in geology and so he was unable to rid himself completely of teaching geology.

The 'good John Young' (1823-1900) had struggled through appalling poverty and hard work - 16 hours a day in a mill - and used his meagre spare time to study the Carboniferous formations near his native Lennoxtown and establish a reputation as an amateur palaeontologist. In 1855 the British Association met in Glasgow and he was given leave from his work to arrange for the meeting a highly successful display of rocks and fossils from the west of Scotland. Rogers and William Thomson recognised his abilities and got him appointed in 1859 as Assistant Keeper in the Hunterian Museum under the Keeper, Rogers. It was the good John Young who undertook the massive task of moving the Hunterian Collections from the old Hunterian Museum at the rear of the High Street College buildings when the University moved to Gilmorehill in 1870. Although lacking a degree he achieved fame as a Carboniferous palaeontologist and was, for instance, awarded the Murchison Fund by the Geological Society of London in 1883 by which time he had authored nearly 50 papers. Eventually in 1893 he was justly (if tardily) awarded an honorary LL.D. by the University of Glasgow. There is no doubt that much of the good order, reputation of the Museum and of the quality of the research and teaching in geology in the University in the years 1860-1900 (when he resigned) was due to the 'good John Young'. His teaching abilities were recognised by part-time appointment to a geological Lectureship in the Mechanics Institution in 1874-81 and then in Anderson's College from 1882, a source of income that supplemented his meagre salary (£40 and free lodgings) from the University; indeed he took a cut in income when he became Assistant Keeper in order to pursue science. He was so well-loved that his friends subscribed to buy him a life Fellowship in the Geological Society of London in 1874. His main researches were on Carboniferous fossils and in conjunction with James Armstrong published in 1871 *The Carboniferous fossils of the west of Scotland* and then in 1876, for the British Association's meeting in Glasgow, the *Handbook of the western Scottish Fossils* - a definitive study. He also published many papers on the local Carboniferous geology.

Undoubtedly the best known Glasgow contributions to 19th century geological thinking (and contemporaneous with both John Youngs) stemmed from William Thomson, later Lord Kelvin, Professor of Natural Philosophy from 1846 to 1899 and President of the Glasgow Geological Society for 21 years (1873-1893). His arguments for the age of the Earth, based on the time taken for it to cool (before heat-producing radioactivity was known) ranged initially from 20 to 500 million years, later 20 to 40 million years. This created great international controversy. Kelvin also contributed to the study of the causes of sea-level variation, glaciers, tides, the age of the ocean basins and particularly to showing that the Earth did not consist of a solid shell enclosing a central mass of molten rock but rather that the centre was more like lead in a press and the Earth as a whole acted in a rigid way. He also recognised the dependence of climate on ocean circulation so that if the distribution of land allowed tropical warm water to flow unrestrictedly to the Poles there would be no ice caps at the Poles. Kelvin's views on the age of the Earth became particularly controversial when it was appreciated that Charles Darwin's view on evolution required lengths of time undreamed of by many of his contemporaries. Kelvin provided an overall theoretical basis to many geological processes while most of the geologists at that time were concentrating on describing and discovering the rock and fossil record without many thoughts on global issues.

The reorganised Hunterian Museum re-opened to the public in 1877 when the charge for entry was sixpence, a substantial sum in those days but only half the twelve pence which in 1844 the old College Museum charged. Admission continued to be free to matriculated students (and two friends each) and the displays were used for teaching. Yet another J. Young, one of the first B.Sc. graduates, was engaged in revising the catalogue in 1878-9.

In the last decades of the 19th century, geology was increasingly regarded as a separate subject with the B.Sc. degree offering three alternative courses of study, in Biological Science, in Geological Science and in Engineering Science. The funding of the Honeyman-Gillespie Lectureship in Geology by Mrs Honeyman-Gillespie in 1876 in memory of her husband was related to still another J. Young - James 'Paraffin' Young - who had established a plant at Bathgate to produce oil by distillation from cannel coal or torbanite. In an 1854 lawsuit, Honeyman-Gillespie maintained that the lease he had let for the working of coal on some land at Torbanehill, near Bathgate, did not extend to highly profitable deposits of oil-rich shale, torbanite, which he maintained were not coal. Unfortunately he lost the case, much to his cost and indignation and he felt that clearer teaching of geology in the University would inculcate the 'correct' understanding of this problem. Another separate geological bequest established the William Baxter Demonstratorship in Geology in 1888. A number of geologically-based student prizes existed by this time and in 1878-9 students were graduating with B.Sc. in Geological Science while in the same year departmental certificates in geology were awarded, among others, to Reisaku Masuda and Naosada Taniguchi being among the first Japanese students to study in the University. But there was still no formal department of geology and serious deficiencies in the teaching of mineralogy, petrology and practical mining existed which were partly filled by , eg, Dr Dobbie who in 1882-3 lectured in mineralogy or eg, Mr G. R. Thompson who from 1896-1900 (at least) lectured in geology. The problems which Professor Young faced both in geology and in the Hunterian Museum and the problems in Science generally were summarised by him in his evidence to the Universities (Scotland) Commission on 24 June 1876 which was published in full.

Professor John Young's hopes of being relieved of all geology teaching were raised when funds became available in the early 1890s to establish a new Chair but much to his annoyance a Chair of History was established with the first appointment in 1893, the year the Faculty of Science was established by the University Commission's Ordinance 31. Any new impetus in geology had to await the death of John Young M.D. in 1902.

The Regius Chair was then renamed for Zoology and Graham Kerr appointed while a new Chair of Geology was established by an Ordinance approved by Edward VII on 11 August 1903. The endowment of £15,000 was provided by the Carnegie Trust, the Bellahouston Trust and others. Thus a separate Department of Geology at last came into being with the appointment in 1904 of the dynamic John Walter Gregory, already F.R.S. and subsequently to be President of the Geological Society (1928-30).

Professor Gregory (1864-1932) who was supported by the influential J. W. Judd, T. G. Bonney & E. Suess, was chosen without interview, being in Australia. He was a superb appointment, being an excellent teacher, an outstanding administrator, a most productive researcher - he published 20 books and over 300 papers - and a very well-liked person. He had obtained a London University B.Sc. in Science in 1891 after studying part-time at Birkbeck Institute while working first as a clerk at wool sales and then, from 1887, as an Assistant in the Geology Department of the Natural History Museum. He rapidly established a highly professional training in Geology in Glasgow, being enormously helped by his previous experience (1900-4) as Professor of Geology and Mineralogy at Melbourne University. Here he had established the Department from the beginning including overseeing the building of laboratories and the obtaining of equipment. In two years in Melbourne he had made geology one of the most important and popular subjects in the University; he played a major role in organising the School of Mining Engineering there and moreover from 1902, with unbelievable energy had also undertaken, in addition to his University work, the Directorship and entire reorganisation of the Geological Survey of Victoria for the State Government. He was a powerhouse of irrepressible energy, a 5mph walker, encyclopaedic in his geological experience, for he travelled extensively - Rocky Mountains and Great Basin of the USA; East Africa, recognising and naming the Great Rift Valley (the first use of the term rift valley) which now bears his name; Spitzbergen, where he made the first crossing; West Indies, Australia, Tasmania, New Zealand, Libya, Angola; Swedish Lapland, Spain, British Columbia, Burma, Tibet and India. His field experience, bravery, toughness, love of adventure, and his ability to work under the most adverse conditions were phenomenal. By repute he could at one time "nurse his infant on his knee, correct the proofs of one book with his left hand while writing another with his right and dominate a polemical discussion on any topic". Much of his writing was accomplished while travelling, for instance, after his fourth visit to Australia he returned via the Pacific Ocean, across Siberia to Moscow and during this journey he wrote a book on the geography of *Australia* (1916). He would sit in a crowded, noisy station, or in a railway carriage or even on the crowded deck of a boat and, oblivious of interruption, write (until the deck chair was swept overboard on one occasion).

He arranged lectures at 4.30pm for the first year and 5.30 for the second year to allow himself time for more demanding activities in the fresher parts of the day (lectures started at 8am in those days) and he regarded the somnolence of Senate meetings as the best time to clear off arrears of correspondence. As one 92 year old ex-student, Miss M. Filbin, wrote in 1992 about the period 1918-21, Professor Gregory "was always rushing". He finished one lecture and raced out only to reappear seconds later: "Had I a hat or have I left it?" - "It's on your head, sir". His lectures were enormously popular because of the wide personal experience he could bring to both academic and practical aspects. He rapidly established goniometer work, blow-pipe analysis, microscope work and, as in the last decades of the 19th century, geology was being taught to engineering students (building stones, road meals, water supply) to agricultural students (rocks and soils, subsoils) to public health students (rural hygiene, sanitary requirements of cemeteries, geology of water supply) to mining and mining engineering students in addition to the subsidiary teaching of Arts and Science students. Additional, of course, was the training of geologists. Field excursions for the Ordinary class involved hiring a train and taking 300-400 students to places such as Campsie Glen where the students would be strung out over hundreds of metres with Gregory in the front like a steam engine and an assistant at the rear. From the beginning the classes were mixed, male and female, as Gregory made no distinction; to an administrator who asked if he would run separate field classes for females he replied that he only had two ladies in the class, one he knew had a doctor's exemption from field work and one was very pretty, but he would need to consult the records to find out which was which before giving an answer. The question of separate field classes was quietly dropped. In 1906 Helen Maxwell Armour was the first woman to be awarded a B.Sc. (Pure Sci.) in Geology, one of a class of seven. The first B.Sc. Honours in Geology was Thomas McCrirck in 1903, just before Gregory came to Glasgow.

Gregory was a polymathic geologist, versatile and familiar with every branch of geology. Thus he published widely in palaeontology eg, fossil bryozoa, echinoids corals, vertebrates, in glacial geology eg, *The nature and origin of Fiords* (1913), in economic geology eg, world resources of iron ore 1907, *Elements of Economic Geology* (1927), in stratigraphy and structure eg, *The structure of Asia* (1929), on the geology of the moon, the origin of artesian water in *The dead heart of Australia* (1906) on climate, on sea-level and reaching an intellectual apex in two presidential addresses to the Geological Society of London on the geological history of the Atlantic (1929) and Pacific (1930) Oceans - a broad scenario of two thirds of the Earth. His last books were *Dalradian Geology* (1931) and *The story of the Road* (1931). He was to foreign geologists, the most prominent and best-known British geologist of his time, an outstanding authority on the world as a whole who rapidly (everything Gregory did was speedy) brought fame and distinction to the Glasgow Department whilst remaining modest and very well-liked by his staff and students.

As would be expected, many of his views turned out to be incorrect and generally it can be said that he was too quick and not patient or thoughtful enough. Thus his rifting origin for fiords, his interpretation of the structures in the Dalradian rocks, of the origin of artesian water in central Australia, and of the origin of the Atlantic as a rift trough have not stood the test of time. His opposition to continental drift was of course the generally held view at that time although he was quick to notice surprising similarities on opposite sides of both the Atlantic and Pacific Oceans. In view of his knowledge of world geology, his opposition to continental drift was surprising and unfortunately very influential. If the physicists said that it was impossible and the geologist who had seen more world geology than anyone else opposed it, who would dare to support Wegener?

In 1914 the external examiner, Professor J. W. Judd C.B., F.R.S., reporting to the Principal on the three years 1912-14 wrote "I have been greatly impressed by the results attained both in written answers and in the skill and knowledge shown in dealing with tests of practical training . . . The teaching has been both inspired and thorough. Glasgow University by its unique position affords greater facilities for giving practical instruction in the field in almost all departments of geological science than the teaching centres of almost any other country and it is gratifying to find these splendid opportunities are being made use of to the fullest extent".

In 1906, shortly after Gregory came to Glasgow, he appointed and personally paid George Walter Tyrrell (1883-1961), a 1904 graduate under Professor J. W. Judd of the Royal College of Science, London (now Imperial College), to be his 'Assistant'. The present departmental pattern of professors and lecturers developed out of Ordinance 17 of 1889 which enabled the University to appoint lecturers "whose teaching shall qualify for graduation" as distinct from 'assistants' who were paid by and employed by professors. Tyrrell became the first lecturer in the Department in 1913 and subsequently also the first senior lecturer, although he was employed in teaching from 1906.

Without G. W. Tyrrell, who subsequently became F.R.S.E., Ph.D. (by special dispensation of the Senate in 1920 in view of his F.RS.E., F.G.S. and First Class A.R.Sc.) and then D.Sc., Gregory could not have undertaken such extensive travelling. Tyrrell was not only a most assiduous and systematic teacher, able to cover a great part of the field of geology, but he was also very well liked, kindly, generous, courteous and thoughtful of others but rather withdrawn and shy, he became a most distinguished igneous petrologist of international renown. Tyrrell visited Spitzbergen in 1919 and 1920, Iceland in 1924 and Siberia in 1937. The bulk of his work was in igneous petrology, much in the west of Scotland eg, lavas and intrusions in Ayrshire, the Clyde Plateau, the Midland Valley and Arran. He discovered the composite alkaline sill of Lugar and described it minutely (1912, 1917, 1948, 1951); he also revised the *Geological Survey Memoir of Arran* (1928), recently after 60 years reprinted in facsimile by the Survey as still the definitive account an incredible achievement, but he also described rocks collected by himself or others from Spitzbergen, Iceland, Greenland, Jan Mayen, Rockall, Antarctica, South America and Africa. His textbook *The principles of Petrology* was first published in 1926 and went through nine editions up to 1948 and was still in print in the 1980s - a remarkable testimony to the careful choice and succinct presentation of basic petrology. Incredibly, for 40 years (1914 to 1953), until crippled by a stroke in November 1952, he reviewed the progress of geology in *Science Progress.* Glasgow's reputation for igneous petrology declined with Tyrrell's demise and has not subsequently regained its previous heights.

It was not until after World War I that the Department really began to expand in response to the greatly increased numbers of students.

Ethel D. Currie D.Sc., Ph.D., F.R.S.E., F.M.A. (1899-1963) graduated under Gregory in 1920 and after a short period as Demonstrator was appointed Assistant Curator of the geological collections in the Museum and eventually senior lecturer in 1960. She lavished care on the collections, in cataloguing, in display and in teaching and her publications were of exceptional quality. Fossil echinoids from Persia (jointly with Gregory) were followed by echinoids and other fossil descriptions from Cyprus, Somali land, Ethiopia, Kenya and Burma; xiphosaurs from Lesmahagow, Jurassic ammonites, and her magnum opus on Scottish Carboniferous goniatites which gave precision to the zonation using these fossils, demonstrated a great break in the Namurian series and showed the Visean age of the Scottish Calciferous Sandstones. For over 60 years the main geological exhibits in the Hunterian Museum stemmed from, or were, her work and some were only recently taken down during the complete revision of Museum exhibits in 1992.

John Weir (1896-1978), who had been three times severely wounded in World War I, was appointed in 1923. He took an onerous and most conscientious role in teaching palaeontology and stratigraphy. He published on Mesozoic and Cenozoic mollusca from East Africa, Spitzbergen and Burma - echoing Gregory's influence, on the Jurassic boulder beds of east Sutherland in combination with Bailey, on non-marine bivalves in the Coal Measures, partly alone and partly with Trueman including a series of monographs on the latter fossils but he also completed a comprehensive memoir on the Belerophontidae of the British and Belgian Carboniferous. He was therefore an outstanding palaeontologist who with Ethel Currie gave strength to this aspect of the Department's research.

The build up of staff continued in the 1920s with the appointment of W. J. McCallien as Baxter Demonstrator and later Carnegie Teaching Fellow with research interests in metamorphic and structural geology; Benjamin H. Barrett as Carnegie Teaching Fellow (in 1924) and subsequently lecturer; Mary H. Latham as Demonstrator and then librarian who retired in 1959, and W. R. Flett B.Sc., initially as Demonstrator, but later as Assistant, before joining the Royal Technical College in 1947.

Gregory also appointed, and presumably personally paid, the first technician in the new Department when in 1906 Arthur Ian Curlette was employed as 'laboratory attendant'. Curlett remained in the Department, apart from a period (1914-19) during World War I, until December 1938 in Trueman's time and Mrs Catherine Maclean who started in 1915 during the War, continued until after World War II. In 1925 Andrew R. Ferguson was appointed at the age of 14 and continued until 1972, retiring as Chief Technician, although the term technician was not generally used until World War II. Ferguson and Maclean saw the Department through World War II. They made most of the thin sections needed by the practical classes and the researchers, in particular Tyrrell, whose work was dependent upon thin sections. These technicians, with a succession of messenger boys, and Jack W. Keith who transferred from Zoology in 1934 but was away during the War, sustained the Department in its first four decades.

So by the time Gregory 'retired' in 1929 he had built up a very large Department compared with most other British Geology departments, with a first year class reputedly the largest in the UK - over 400. Gregory resigned his Chair, not to retire, but to have more time for writing - he published four books, two presidential addresses and 30 papers in 1929-32 - for travel and to act as President of the Geological Society of London. He was drowned when a canoe overturned in Peru in 1932.

Gregory's successor was Edward Battersby Bailey (1881-1965) another dynamic geologist already with an international reputation and a man who bubbled with enthusiasm for field geology.

Bailey was an extrovert, a sportsman, keen on boxing and arduous fell walking, able to sleep outside in the open and a man who had consciously toughened himself to be able to undertake rigorous fieldwork in the highlands under any conditions and enjoy it. He was a Cambridge graduate in Geology and Physics who had entered the Geological Survey of Scotland immediately after graduating in 1902. He had been trained initially by the legendary geological surveyor, Ben Peach F.R.S. and then by that self-effacing genius C. T. Clough. He mapped in the Scottish lowlands and highlands before 1914, produced with others, the classic synthesis on the Glen Coe cauldron subsidence, the first ancient (400 million years) caldera subsidence to be recognised, and from 1907, until the summer of 1914, was engaged with others in the Survey of Mull, which revealed two great centres of cauldron subsidence. He became increasingly interested in the structure of the southwest highlands, the complexity of which defied unravelling by others. He volunteered for action in World War I and was reported to be fearless, almost reckless, in the face of danger. He gained a Military Cross but was seriously wounded twice and lost one eye, damaged one arm and had other wounds. After the War he returned to the Geological Survey, completed the mapping of Mull which eventually resulted in what is generally agreed to be the finest colour printed 1:63,360 Geological Survey map ever produced - full of fine detail. Out of this work came nascent ideas on magma types. Bailey found himself however increasingly at loggerheads with the senior Survey staff because of his unconventional attitudes. For instance, he insisted on always wearing shorts in the field at a time when the gentlemen of the Survey wore plus fours. He regularly walked into rivers early in the day so that he need never worry about soaking his feet. Once he led an assistant to the Director through a highland river chin-high although there was a serviceable bridge only a few yards upstream. He was fired with structural interpretations and relationships within the Survey became further strained when he re-interpreted the mapping of other Survey officers to his own views. Eventually the Director banned him from being allowed to examine the field maps of sheets that Bailey had not mapped himself! When, therefore, the Chair of Geology became vacant in 1929, Bailey applied and was appointed from January 1930, the same year as he became F.R.S. Thus the first two Geology professors were not only highly distinguished even before appointment, but they set a firm seal on value of unflagging, vigorous field work and wide-ranging structural syntheses with total disregard for adverse circumstances and trivial - or even sometimes important - detail. They established the Department as one which turned out reliable, competent, undaunted field geologists.

Bailey's outstanding contribution was the recognition and formulation of the nappe synthesis of the structure of the Dalradian rocks and the synthesis of the tectonic structure of the southwest and central highlands of Scotland. In part this stemmed from a 1909 visit to the pre-alps of Switzerland with L.W. Collet, in part from the mapping and re-mapping of the Dalradian rocks, for instance at Ballachulish-Fort William and in the central highlands. Bailey's tectonic syntheses had been severely handicapped by not being able to objectively determine the stratigraphic sequence in the Dalradian rocks ie, which were the oldest and which were the youngest rocks. Shortly before he applied for the Glasgow Chair Bailey was shown by T. Vogt how to use graded bedding and cross-bedding to determine whether lithological successions were inverted or right-way-up relative to their depositional positions. Suddenly it became possible to determine the original sequence in the Dalradian rocks and to determine which folds were anticlines and which were synclines. Bailey immediately (1930) set a young Glasgow graduate, Archibald Allison, to establish the sequence in the Islay, Jura and Loch Awe regions and from the results published jointly, Bailey was able to go on to revise his syntheses of the overall structure of Dalradian rocks in 1934 with nappes which had moved to the northwest and southeast with the steep Loch Awe synclinorium structure in between. It was

Bailey who introduced the term ' to young' (1934) and, with W. J. McCallien in 1937, the terms antiform and synform. It is ironical that as Bailey and Allison were establishing what is now accepted as the true succession in the Dalradian, Gregory was publishing his quite different and incorrect view.

Following Bailey's 1934 Dalradian synthesis he produced a collection of *Tectonic essays mainly Alpine* (1935) which expanded on the nappe structure of the Alps and then with O. Holtedahl a review of the whole of the *Caledonides of NW Europe* (1938). From initial purely objective mapping Bailey became more and more involved in interpretative synthesis so that Dr Richey, an old Survey colleague wrote "his first objective was to achieve a basic pattern, and from that to fit in other data. The guiding idea, once gained, either by arduous work or more usually by a kind of intuition, became his ruling passion". This of course ran contrary to the Survey ethic of completely objective mapping.

Bailey was an interesting lecturer, though not very systematic in adhering to a syllabus. He was difficult to stop and lacked an appreciation of what could be fitted into the available time. One evening lecture, started at 7pm, was continuing after 10pm when he realised the audience had dwindled to two - the secretary who had arranged the lecture and the janitor who was waiting to lock up. Bailey turned to the janitor and asked if it was alright to continue as he was coming to the most important part of the lecture, to which the janitor replied "Go on as long as you wish, I get overtime after two hours". To the students he was a hero and his vast fund of geological experiences kept his lectures full of interest. Like Gregory before him, field work was an arduous experience, often carried out in rain, hail or snow with no thought for discomfort - or inconvenience. On one occasion he arrived at a quarry with 300 first year students only to find that a row of terrace houses had been built in front of the quarry since his last visit. Without pause he marched up to the front door of the middle house and innocently asked the surprised housewife if he could take his students to see the quarry. After receiving her assent, 300 booted students tramped in through the front door and out through the back and then repeated the process in reverse after looking at the quarry, much to the chagrin of the housewife.

Bailey's time at Glasgow was extremely fruitful but comparatively short and in 1937 when the Directorship of the Geological Survey fell vacant, Bailey accepted the post and was eventually (1945) knighted, awarded the Wollaston Medal (1948) and seven honorary degrees. He continued to write and carry out fieldwork until the early 1960s and the writer recalls Bailey vigorously climbing Bray Head in Ireland (in shorts) in 1956 when he would have been 75. Bailey and

McCallien collaborated most profitably with for instance, joint papers on the Dalradian rocks of northeast Antrim (1934) and central Perthshire, Schiehallion to Glen Lyon (1937), and even long after both had left Glasgow and McCallien was in Ghana, on Serpentine lavas, the Ankara melange and the Anatolian thrust in central Turkey, the latter being of major importance in the interpretation of the tectonics of central Turkey. Nevertheless, Bailey's seven years in Glasgow were not long enough, or supported with enough co-workers, to give Glasgow a strong reputation in tectonics. Glasgow was seen as an interlude for Bailey in what was essentially a Survey career.

Bailey's successor, Arthur Elijah Trueman D.Sc. (1894-1956) previously lecturer in Cardiff (1917-20), Head and then Professor of Geology and Geography at Swansea (1920-33) and then Professor of Geology at Bristol (1933-7) was, like Gregory and Bailey before him, already distinguished on appointment, but in most other ways he was an entirely different man, weaker in constitution, less able to sustain field work and essentially dedicated to palaeontology and landscape evolution. He had made a name for himself initially in studying Jurassic (especially Liassic) ammonites, gastropods and lamellibranchs, particularly evolutionary changes in *Gryphaea* and then, deriving from his appointments in south Wales, he had become a world expert in Coal Measure non-marine lamellibranchs with which he was able to zone the economically important Coal Measures in a detail not possible before. With John Weir and Duncan Leitch (a 1926 Glasgow graduate who stayed on in junior teaching positions and was made a lecturer in 1938) already researching and publishing on Coal Measure non-marine lamellibranchs and Ethel Currie's work on Carboniferous marine goniatites, an outstanding and distinctive Glasgow school of palaeontology was established. The work was both academic in identifying evolutionary changes and in studying statistical variations within species, and of profound economic importance in tracing and identifying coal seams, especially in the highly fractured coal fields of Britain, but also of course, internationally.

Trueman became F.R.S. in 1942, largely for this work, and it was he who carried the Department through World War II. He resigned in 1946 to become full-time Deputy-Chairman and then Chairman of the University Grants Committee until forced by ill-health to resign in 1953. He was President of the Geological Society from 1945-47, knighted in 1951, awarded the Wollaston Medal of the Geological Society in 1955 and had received four honorary LL.D.s by the time of his death in 1956. His last book, *The Coalfields of Great Britain* (1954) edited and partly written by him, was a fitting culmination of his life's work but was not as popular as his well-known *Scenery of England and Wales* (1938) which he had largely illustrated with his own drawings. Trueman was responsible for initiating statistical studies of variation in fossils and is remembered for his perspectives of evolutionary principles and the meaning of observed variations; arid description without carefully considered interpretation was anathema to him. He was an outstanding lecturer, lucid, terse, cogent, witty and well-timed. He had qualified as a teacher and retained a life-long interest in education, including adult education; he published books on geology for school-children and endeavoured to popularise science.

Trueman appointed Dr A. T. J. Dollar in 1939 but Dollar was away in the War until 1946. Dollar is mainly remembered for his work on earthquake monitoring but while in Glasgow he also visited and studied the volcanoes of Jan Mayen in 1947 before leaving in 1950 for Birkbeck College, London.

Under Trueman's influence and because of the staff left to him during World War I, the Geology course became more palaeontologically oriented, a process continued by his successor and therefore effective for 37 years in total.

Trueman's successor was Professor Thomas Neville George D.Sc. (1904-1980) subsequently FRS (1963), and Hon. D.-es-Sc. Rennes (1956). In a factually correct way, he was almost a Trueman-clone. He had been trained as an undergraduate by Trueman, entering University College at the same time as Trueman, when the College opened in 1920; he had done research and taught under Trueman, he had absorbed Trueman's interests in Jurassic and Carboniferous palaeontology and Carboniferous stratigraphy, also his interest and appreciation of landscape evolution, likewise his ability to teach both geology and geography, his absorbing concern in, and deeply felt obligation to promote, adult education not merely in geology but widely in science, and Trueman's ability to give an absorbing pedagogical lecture; both men could think on their feet and express themselves in well-structured English.

Apart from a brief period of two years at Cambridge as a Ph.D. student under Mr Henry Woods F.R.S. - a dry palaeontologist given to meticulous morphological description devoid of interpretation, still less speculation on evolutionary principles - and three years in the Geological Survey, George had spent 38 of his first 43 years in Swansea and had been Trueman's successor to the Chair in Swansea. Where George differed from Trueman was his period with the Geological Survey during which he became an expert on the geology of Wales and also a good field geologist, active and robust. Unlike Trueman, he loved to provoke an argument, usually over an intellectual matter, but sometimes over the distribution of resources or the arrangements within the Hunterian Museum where the shared responsibilities and lack of an overall director (until 1976) left fertile fields for disagreements.

George published about 150 papers and another 100 reviews, contributions to discussions etc, mainly about Carboniferous stratigraphy and palaeontology, especially concerned with south Wales and the Welsh borders but also including Ireland and to only a small extent Scotland, apart from a masterly review of the origin of the midland valley of Scotland given in a centennial presidential address to the Glasgow Geological Society in 1958. His best known and most enduring publications were two regional geology handbooks written for the Geological Survey, *North Wales* (with B. Smith), first published 1935 and *South Wales* (with J. Pringle) first published in 1938 and both went through several revisions and editions by George over the next 40 years. These regional geology handbooks exemplified George's gift of synthesis and instruction but inevitably in a pre-plate tectonic context for he was over sixty when plate tectonics became generally accepted. George was fascinated by the palaeogeography and depositional environments in the lower Carboniferous and he repeatedly returned to this theme; in later years he became embroiled in controversy over the reality of eustatic movements in Carboniferous times.

George also became involved in prolonged tussles with his staff as the old concept of the Department as a professorial kingdom crumbled away - a concept emphasised in the above account by the concentration on the role of the professor. He was also renowned for his parsimony in Departmental expenditure but as a scientist he was a most assiduous worker, unflagging in his output even after retirement in 1974 at 70 and his last major paper, over 50 pages long, published in 1978 when he was 74, returned to the deposition of lower Carboniferous limestone in Gower, south Wales.

George's successor in 1974 was Bernard Elgey Leake, D.Sc., Head of Department 1974-1992, a metamorphic and igneous petrologist with strong geochemical and mineralogical interests and devoted to unravelling the geology of Connemara, western Ireland.

Despite the large numbers in the Ordinary (first year) class and the reputation of the first professors, relatively few alumni became geologists of distinction until after World War II when Professor George estimated that from 1946-76, 15 members of the Department had become university professors and six directors of national Geological Surveys. Three of the earliest of Gregory's students to achieve outstanding success were Sir James Mann Wordie C.B.E.

(graduated 1910), the polar explorer and geologist, James Phemister (graduated 1915), pioneer geophysicist and eventually Petrographer and Assistant Director of the Geological Survey, and finally Ven Kiang Ting (graduated 1911), founder and first Director of the Geological Survey of China - a vast organisation, perhaps the largest Geological Survey in the world. Among the first graduates and research students, post-World War I, were W. Q. Kennedy, later Professor of Geology at Leeds, F.R.S. and Director of the Institute of African Geology at Leeds; T. C. Phemister, who went on to study the Batholith of British Columbia and became Professor at Aberdeen and W. E. Swinton, later an outstanding palaeontologist at the Natural History Museum. J. V. Harrison who later achieved fame for work in Peru, Borneo, Colombia, Honduras, Mexico, Venezuela, Trinidad and most of all Persia, had all the stamp of Gregory's exploration enthusiasm. Murray MacGregor and Colin Leitch rose high in the Geological Survey, and Charles S. Pichamuthu of Bangalore (Ph.D. 1937) became a world renowned expert on the Precambrian rocks of Mysore, especially charnockites, Director of the Mysore Survey and also held several professorial posts. In general however, probably more alumni entered the oil industry than any other occupation and many rose to high positions (eg, A. Allison, J. Spence, D. Harkin and I. Vann).

Among the staff to come after 1940, Dr Norman Holgate, who was appointed by Trueman in 1943, made an important contribution to the recognition of liquid immiscibility in magmas at a time (1954) when this was almost universally disbelieved following N. L. Bowen's 'demonstration' that this process was not significant in natural magmas. Holgate's thought was perceptive, especially his invocation of immiscibility in the Skaergaard rocks, but he did not follow up the work so that it was largely overlooked until two decades later when liquid immiscibility came back into fashion.

Dr Basil C. King, who succeeded Tyrrell in 1948, unravelled, by careful mapping, the caldera collapse mechanism and the hybridization of igneous rocks in the Arran central ring complex of tertiary age and became heavily involved in mapping the Dalradian rocks of the central highlands during his time in Glasgow. He left for the Chair of Geology at Bedford College, London in 1956, returning to his earlier studies of the alkaline magmatic rocks of Uganda for which he is mainly remembered. Much of his Dalradian work around Braemar, was never completed or published.

Geophysics entered the Department in 1948 with the appointment of D. M. Boyd (who left in 1955 for the Rio Tinto Co.) but only became firmly established with the appointment of Dr A. C. McLean in 1954, concentrating on the subsurface geology of the Midland Valley and Clyde Estuary.

In summary, up to 1989, when the Department amalgamated with the Geology and Applied Geology departments of Dundee and Strathclyde, the Departmental research had fallen into a number of distinct fields. Under Bailey, the study of the tectonics and metamorphism of the Dalradian rocks included W. J. McCallien and A. Allison and subsequently J. G. C. Anderson, B. C. King, N. Rast, T C R. Pulvertaft, G. Voll, D. M. Ramsay, D. R. Bowes and A. E. Wright who all made contributions in this field. The focus of attention shifted to the Lewisian basement following the arrival of D. R. Bowes in 1956. He correctly demonstrated the existence of early metamorphosed basic intrusions distinct from the Scourie dykes, identified the existence of a whole series of basic dyke injections rather than one synchronous series of 'Scourie dykes' and put the structure and geochronology of the Lewisian complex on much more secure basis. R. G. Park and A. E. Wright were co-workers in the project. From northwest Scotland, Bowes extended his investigations into Precambrian complexes in USA, Finland, India, USSR, China, South Africa and Brazil. Dalradian investigations continued in Connemara, western Ireland and included A. Rothstein (Dawros Peridotite), B. W. Evans (partial melting near ultrabasic intrusions), B. E. Leake (structure; metamorphism; origin of gneisses granites and amphibolites, geochemical studies) and P. W. G. Tanner (structure and metamorphism). Out of this work came amphibole studies which led B. E. Leake eventually to the first systematic classification of the complex amphibole group and which is now internationally-approved.

A prestigious asset has been the Isotope Geology Unit of the Scottish Research and Reactor Centre, founded in 1966 jointly by the Scottish universities. This has led to a series of front-line international isotopic dating studies which pioneered the use of stable and radiometric isotopes in studying geological processes. The many collaborative projects between Geology departmental researchers and the Isotope Geology Unit staff brought outstanding credit to both groups.

A major field of continuing study in the Department has been that of stratigraphy and palaeontology from E. D. Currie and J. Weir, R. M. C. Eager, W. E. Swinton, D. Leitch, A. Lamont, A. E. Trueman, T. N. George, A. Williams, L. R. Moore, R. H. Cummings, K. A. G. Shiells, R. B. Wilson, O. Dixon, G. M. Dunlop, G. Y. Craig, J. K. Ingham, W. D. I. Rolfe, D. F. B. Palframan, W. W. Bishop, J. D. Lawson, M. C. Keen, C. J. Burton, A. W. Owen, with the first scanning electron microscope being installed in 1976. The last decade saw developments in pioneering molecular palaeontology under G. B. Curry.

In sedimentology E. K. Walton (greywackes), W. G. Jardine (Quaternary deposits), T. N. George and G. E. Farrow (carbonates) and B. J. Bluck (clastics) all established reputations in their own fields with B. J. Bluck becoming a leading authority in Scottish terrane determination using sedimentary deposits.

Geophysical research has involved gravity, magnetic and seismic studies of the midland valley, southern uplands and the Dalradian block mainly involving A. C. McLean, J. Hall, D. Powell and D. R. Watts and their students. Geochemistry was applied by D. R. Bowes and B. E. Leake to the study of sedimentary, metamorphic and igneous rocks and by C. D. Gribble to rock aggregates for engineering purposes with the first X-ray spectrometer being installed in 1968, the first electron microprobe in 1976 and the first automatic X-ray spectrometer in 1978, the last two instruments promoting the introduction of the first significant computing facilities in the Department led by Dr C. M. Farrow.

The one thread that distinguishes almost all these investigations is the lack of theoretical modelling and speculation in favour of empirical observation and derived deduction, ie, 'go and see first' and then interpret and in this field work and collecting has usually been the basis of most of the projects. This was well brought out in the Departmental research papers in the book that celebrated the formal 1977 opening of the Geology building, *Crustal evolution in northwestern Britain and adjacent regions* (1978) edited by D. R. Bowes and B. E. Leake.

In addition to the books mentioned throughout this account, there have been three textbooks published within the last 15 years which have achieved success; *Geology for civil engineers* (1979) by A. C. McLean and C. D. Gribble, *A practical introduction to optical mineralogy* (1985) by C. D. Gribble and A. J. Hall and the 27th Edition of Rutley's *Elements of mineralogy* (1988) revised by C..D. Gribble. All these have been used in the Departmental teaching which remained broad and unspecialised with a particularly high standard of field work training following the standards set in the Department from the beginning.

First year lectures were given in the East Quadrangle in the old Zoology lecture theatre (now the Gregory Room) until the numbers exceeded the lecture theatre capacity and then the new Zoology and later the new Chemistry lecture theatres were used. Andrew Ferguson recalled taking sets of large rock specimens in a wheel-barrow to the Zoology Department for Bailey's lectures. Geology expanded into more of the East Quadrangle accommodation when Surgery moved and the Surgery Museum became a Geology Honours laboratory. The

Practice of Medicine Lecture Room became the Level 2 (Higher Ordinary) Geology Laboratory and first year laboratories were on the ground floor and in the attic. Anatomy adjoined Geology until Geology moved out into the new Geology building in 1976.

Increasingly as research work in Geology became more laboratory oriented and the research workers expanded in numbers, the restrictions of space and lack of suitable laboratories became a serious restraint to the Department's research eg, in geochemistry which from initial work by Dr N. Holgate really began when Dr D. R. Bowes came from Swansea in 1956. Professor George began pressing for a new purpose-built Geology building in the late '50s by which time the Department had acquired an 'annex' in 61 Oakfield Avenue where geophysics in particular was accommodated. Research students were largely in the attic in the East Quadrangle, initially undivided but later subdivided. After repeated unsuccessful attempts, eventually agreement was reached with the University Grants Committee to fund a new Geology building and foundation work started in 1973. The credit for obtaining this must be given entirely to Professor George and this excellent building remains a monument to him although he had retired before the Department moved into the new premises in 1976. Subsequently, following the Earth Science Review of 1987-89, a three floor extension to the back of the building was added in 1991 to provide a rock store, additional space for research students and a conference room. This building is one of the finest assets of the Department and a springboard for future investigations.

The 1987-89 Earth Science Review resulted in the formation of a new and greatly enlarged Department of Geology & Applied Geology in 1989 which incorporated staff from Dundee, Glasgow and Strathclyde universities and appropriately closes this account.

Acknowledgements

The writer acknowledges the substantial help derived from T. N. George's 1975 account of *Geologists at the University of Glasgow* and from numerous published obituary notices. Dr G. P. Durant provided significant archival information and Mr A. Ferguson verbal and written accounts of the Department from 1924.

Bernard E. Leake is Professor of Geology at the University of Glasgow and Graham P. Durant is Curator in Geology at the Hunterian Museum.

Medical Connections

Robert Young Thomson and John Shaw Dunn

The Faculty of Science as constituted in 1893 included the professors not only of Mathematics, Natural Philosophy, Chemistry, Natural History, Astronomy and Botany but also those of Anatomy, Physiology and Engineering. The connection with Engineering was severed 30 years later when an independent Faculty of Engineering was established, but the link with Anatomy and Physiology remained, and was to prove, to an extent that no-one could have foreseen, a source of new initiatives in research and teaching. It will be convenient to distinguish the very different traditions of the two departments, and to take Physiology first.

In 1893 the Regius Professor of Physiology was John Gray McKendrick, who made striking advantages in the study of the special senses, especially vision, and who after his retirement found a second career in local politics in Stonehaven, where he ultimately became Provost. It is interesting that he was the first incumbent of the Chair to devote all of his energies to his academic duties. (His predecessor had practised part-time as both surgeon and physician and owed his university position to the political patronage of the Whig government which had established the Chair). Although nowadays almost a forgotten figure, McKendrick did much to lay down the lines on which his Department would develop long after he retired from his post in 1906. He planned its laboratories in the newly-built West Medical Block; his interest in the special senses led, albeit indirectly, to the establishment, first of a Lectureship in Psychology and then to an independent department; and he persuaded the Principal, a year before his own retirement, that a legacy left to the University by a Dr John Grieve be devoted to establishing a Lectureship in Physiological Chemistry. He had already demonstrated his enthusiasm for this specialty. He had in 1897 issued to his students both in Medicine and in Science a *Syllabus of Physiological Chemistry* which, amended in detail in successive editions, but little changed in scope, substance or doctrine, was used by the present writer as a student 50 years after its first publication. McKendrick's success in directing the Grieve money to this purpose is an interesting story. Grieve was a Glasgow graduate, a medical practitioner who owned substantial property in the centre of the city. His will, originally drafted in 1887, ran to six printed pages, and was amplified by nine further 'holograph writings' written over the nine years preceding his death in 1896. The University was to receive three-sevenths of his estate towards "endowing a lectureship in Natural Science". One of the holograph writings amplified the testator's intentions at length; for present purposes it is sufficient to say that considerable discretionary powers were entrusted to a committee of five academics; the Principal and the professors of Natural History, Botany, Anatomy and Physiology. The outcome is recorded in elegant copperplate in the minute-book of the University Court. McKendrick and the Professor of Anatomy (Cleland) favoured a Lectureship in Physiological Chemistry. The professors of Botany (Bower) and Natural History (Graham Kerr) favoured using the money to finance zoological research. The Principal came down in favour of the Lectureship, and shortly afterward Edward Provan Cathcart was appointed as its first incumbent. A year later, on McKendrick's resignation, Diarmid Noel Paton was appointed his successor. These developments not only determined the future of the Physiology Department for half a century; they were to have repercussions far beyond Glasgow.

The roles of Noel Paton and Cathcart were so central in determining the relationship between medicine and science in Glasgow that some sketch of their personal histories is warranted. Diarmid Noel Paton was the eldest son of Sir Joseph Noel Paton, a painter much esteemed in late Victorian Britain and well-regarded even now. At age 29 he had been appointed Superintendent of the Research Laboratory of the Royal College of Physicians in Edinburgh, which he speedily turned into one of the most productive medical laboratories in Britain. He spent almost 20 years there before coming to Glasgow as Regius Professor. His tenure of that office was marked by the bitter controversy over the causes of rickets and his obituarists were perhaps more diffident than they need have been in defending the views he put forward with an ill-advised but not ill-founded belligerence. But as a teacher, especially in formal lectures, he was clearly superb. A student described his tall, spare figure, his immaculate grey suit, the bravura performance of his introductory lecture, the brilliance (perhaps learned from his father) of his extemporised blackboard sketches, and his sudden, devastating descents into laboratory classes.

E. P. Cathcart, who had been Grieve Lecturer under Paton and subsequently his fellow-professor in the Department, was a very different person. Eighteen years Paton's junior, he had enjoyed a more varied and adventurous education, including several years in Munich and Berlin. Paton's influence had been diffused, broadcast over all his students by the brilliance of his oratory and his manifest passion for his subject. Cathcart disliked lecturing but was a brilliant conversationalist. Paton would come into the laboratory in an immaculately ironed short white coat, Cathcart in his worn army field service jacket with its rank badges dissected off. (Paton on a celebrated occasion mistook Cathcart for a student and rebuked him for idleness.) Paton was, in private life, an aloof figure. One of Cathcart's strengths was his ability to recruit, as Noel Paton did not, a band of distinguished disciples.

Paton and Cathcart had one characteristic in common, an interest in the British diet, and particularly in the relation between diet and social class. These matters had been studied in the mid-19th century, in Scotland and then England by Hutchison of Carlowrie, and in Britain and Ireland as a whole by Edward Smith. Hutchison was a Lothian laird, one of the principal landowners in the parish of Kirkliston, West of Edinburgh, and a prominent figure in the Royal Highland and Agricultural Society. Smith was an emissary of the Privy Council sent to Lancashire, when the American Civil War interrupted the supply of raw cotton, to assess whether the consequent distress might result in the same famine and pestilence that had devastated Ireland in the potato famine of the late 1840s. Both produced reports which still retain great historical value as accounts of what people ate and how adequately they were nourished, but they were limited by the nutritional knowledge of the day. Intake of nitrogen was recognised as necessary for 'flesh-forming'; intake of carbon as a measure of availability of energy.

Noel Paton's first essay in dietary surveys was undertaken while he was still at Edinburgh, but since it marked the beginning of what was to become, and in a degree still remains, a Glasgow tradition, it merits detailed description. Noel Paton had contributed, and continued to contribute, to a wide range of physiological problems, not always successfully: he was described by an obituarist as one of the last "all-round" professors of physiology in this country. But he was also the first British physiologist to devote serious study to the problems of metabolism; and he had a particular interest in diet. In 1900 he had, under the patronage of the Town Council of Edinburgh, published *A Study of the Diet of the Labouring Classes* in the city.

In several respects Noel Paton's study of the Edinburgh labouring classes marked a new departure. Though small in scale it was the first systematic investigation of what people ate since the 1860s; and it was the first to be directed at the urban poor. Noel Paton was fortunate in his collaborators. J. C. Dunlop, who looked after the chemical side of the investigation, had already studied prison diets - the Victorians were anxious that these should be adequate, but no more than adequate, and cheap. Dr Elsie Inglis, who supervised the visitation of the homes of the participants, was a pioneer of medical education for women. She was subsequently to organise an ambulance in Serbia in the First World War and to die prematurely in 1918, worn out by exertion and suffering.

In comparison with his predecessors, Noel Paton enjoyed immense advantages. By about 1900, methods had been devised for the analysis of foodstuffs into carbohydrate, fat, and protein; the energy made available by the combustion of each of these was known; and the energy, and less certainly, the protein requirements of adults and children, male and female, had been assessed. Given the ages and sexes of the members of a family and the amount of food they consumed, it was possible to decide whether their collective energy and protein requirements were being met. Previously, the assessment of the adequacy of a diet had been a matter of guesswork; the Victorian soldier had, at least on paper, more food than he needed; the Victorian convict was maintained on allowance which ensured that his sojourn in prison would not improve his physique.

In his Edinburgh study, Noel Paton applied methods which had been devised and practised in the United States. In essence, a team of social workers studied a group of families for a week, observed the conditions in which they lived, recorded their income and expenditure, measured accurately their food consumption, and calculated its equivalent in terms of carbohydrate, protein and fat, and hence of calories. The object was not to estimate the extent of malnutrition in the city, but rather to find out how the very poor managed to sustain life. The term 'labouring classes' was a precise one. It meant those whose contribution to the economy was only their labour, as opposed to the 'artisan classes' who contributed also their skill.

Noel Paton's report casts light on the social conditions as well as the nutritional status of the families he studied. Their incomes were small, derived chiefly from the father's wage of about 20 shillings a week. This might depend on the weather or the demand for labour. But it might be supplemented by the (much smaller) contributions of the elder children and the occasional earning of the mother. Children were numerous, more than four to a family, and most under ten. They lived, for the most part in two-room flats. On average, 62% of the family income was spent on food and 13% in rent; 5% went to 'societies' to insure against the costs of illness but more especially of funerals. (A respectable funeral might cost ten week's wages.)

The diets are equally interesting, contrasting as they do with the oatmeal and milk which sustained the peasants studied by Hutchison 40 years previously. They unmistakeably reflect modern urban habits. They would be accepted (though not with enthusiasm) by a present-day Scotsman on supplementary benefit. They are based on bread. Rather more than one pound was consumed, per adult male per day. It provided over 40% of both the calories and the protein, and it cost only three half pence. Only one and a half ounces of oatmeal was consumed, though it was marginally cheaper than bread. After bread, the main components were the expensive, protein-rich foods: about five ounces of meat of various sorts, with about one ounce of fish and the same quantity of eggs. Together these provided 15% of the calories and 30% of the protein. But they cost only threepence halfpenny. The third main source of calories (but nothing else) was sugar: three ounces were consumed daily, together with one ounce of syrup and jam. It contributed 14% of the calories. Since the sugar cost only a farthing it was even cheaper than bread as a source of energy. Butter, about one ounce, provided only 6% of the calories. It cost a halfpenny and as an energy source was more expensive than bread, but cheaper than meat, fish and eggs. It was hardly ever replaced by margarine. Potatoes, about half a pound a day, contributed only 5% of the calories, perhaps because at a cost of over a halfpenny they were, as a source of energy, twice as expensive as bread. Noel Paton did however recognise their value in preventing scurvy in a diet devoid of fresh fruit. Rather more than one third of a pint of milk was drunk, contributing a further 5% of the calories, and costing three farthings. But very little cheese was consumed (half an ounce) no beans, few peas and no fresh fruit.

On the face of it, this diet is not conspicuously deficient in quality or quantity, though Noel Paton thought it provided less energy and less protein than were needed for health, an opinion he later retracted. He pointed out, quite correctly, that better value for money could be obtained by replacing some of the bread and butter with the traditional porridge and milk, supplemented by peas and beans. He speculated interestingly, if unkindly, on the disuse of porridge. In the old days the housewife had to make her own bread as well as her own porridge. Nowadays she purchased bread from the baker and "it becomes a question of using the ready cooked bread or the uncooked oatmeal, laziness decides which it is to be, and the whole family suffers". This was unhistorical as well as uncharitable. In 'the old days' the housewife did not make her own bread; breadmaking requires an oven, a rare luxury. Traditionally the Scots ate scones or bannocks, baked on a girdle; and as for laziness, oatmeal is converted to brose simply by pouring boiling water over it.

In 1911-12, a few years after his appointment to the Chair, Noel Paton instigated a parallel study in Glasgow conducted by Miss Dorothy E. Lindsay. This was broadly on the same lines as its predecessor and it yielded strikingly similar results. But it was larger in scope, covering 60 households instead of 16. This allowed a correlation to be made between income and diet, from which it became clear that, while a regular income of over 20 shillings a week was sufficient to secure a diet approaching the proper standard for an active life, those with a smaller income, and those with an irregular income, entirely failed to get a supply of food sufficient for the proper development and growth of the body, or for the maintenance of a capacity for active work. In one other important respect the Glasgow study broke new ground. In no less that 14 of the 60 families studied, one or more of the children suffered from rickets. The diets of such 'rickety' families were compared with those of the 'non-rickety' families. In general, they provided fewer calories, less protein and, in particular, less fat; but the correlation was not very dramatic. Moreover in the 'rickety' families there were an average 3.9 people to a room and in the 'non-rickety' only 2.9. The conclusion drawn was that "probably both factors - the lack of fat in the diet and the want of sufficient air space - are to be held responsible for the prevalence of this disease among the poorer labouring classes".

Rickets was unquestionably a major public health problem in Glasgow in Noel Paton's time and even later. One can still see today elderly men and women showing the bow legs which were the classical manifestation of the disease. And Murdoch Cameron, the eminent Glasgow obstetrician, had just reinvented Caesarian section to make childbirth feasible for rachitic mothers whose distorted pelvises made normal delivery impossible. The cause of the disease was unknown, and Noel Paton, with Leonard Findlay, an eminent Glasgow paediatrician, sought to identify it by an epidemiological study. Their work was a model of its kind. A great variety of factors was examined: whether the parents were town or country bred; how long the children had been breast-fed; whether they had sufficiency of fat during infancy; whether they had a sufficiency of milk and butter. The results were clear-cut; the diets of rachitic and non-rachitic families were not very different, though the latter had about 11% more fat. But the striking correlations were with the cleanliness of the household and the extent to which the child was taken out into the open air. Rickets was an environmental disease. What mattered in the environment remained ambiguous, but it was not unreasonable for those familiar with the overcrowding and the cloacal atmosphere of the tenements to think that fresh air and exercise were all-important. Had Noel Paton and Findlay rested on these observations they would have been wise. But they were drawn into an acrimonious dispute with Edward Mellanby, who believed that rickets was a dietary deficiency disease. Since it could demonstrably be prevented and cured by cod-liver oil, the Glasgow workers had much the worse of the argument. It is only in recent years, when the importance of vitamin D synthesis in skin exposed to sunlight has been measured, that Noel Paton and Findlay have been, in great part, vindicated. But in any case, Noel Paton, from doubting the existence of vitamin D, had gone on to question the validity of the whole of what he insisted on calling "the vitamine hypothesis". By so doing he greatly damaged his standing. Sixty years later Robert Garry, who had been a student when Noel Paton was Professor, was still appalled at his lack of judgment.

Though Noel Paton was, and perhaps still is, harshly judged, his successor was undoubtedly more successful as an experimentalist and more influential in developing the discipline he professed. Cathcart's interest ranged widely. As a young man he had worked with Pavlov: the last paper to bear his name concerned the innervation of the nipple. But his primary concern was with the application to human physiology of the limited range of chemical methods available for its investigation. The range was indeed very limited. The complex electronic equipment and wide range of reagents on which modern biochemistry relies lay far in the future. The analytical equipment available was that found in any contemporary chemical laboratory; balances, pipettes, burettes and primitive colorimeters; with perhaps a machine for blood gas analysis. The methods of elementary analysis were those of classical organic chemistry. The determination of such compounds as glucose and creatinine, urea and ammonia was a laborious undertaking requiring skill and experience if reliable results were to be obtained. The uptake of oxygen, the output of CO_2 could be measured. The intake of food, loss in faeces and output in urine could all be analysed. But little else could be attempted. Nevertheless, in the hands of an expert such as Cathcart, experiments on these lines were made to yield a great deal of important information. The papers in which it was embodied, written with clarity and simplicity and with a literary elegance that no modern editor would permit, are worth reading even today.

A notable example is Cathcart's first major venture on his arrival in Glasgow, a study of the metabolic effects of fasting. He measured his subject's loss of weight: initially rapid (two pounds a day) but gradually slowing down (three-quarters of a pound a day after a fortnight). The pulse rate fell, like the body weight, markedly at first, more slowly thereafter. Kept in a room his attendants found uncomfortably warm, the starving man complained bitterly of cold. No bowel movements occurred, so Cathcart was spared the disagreeable task of analysing faeces. But his meticulous analysis of each day's output of urine revealed the breakdown of body protein and consequent release of cell constituents, such as potassium, which developed within 48 hours of the last meal and continued, though at a decreasing rate, throughout the fast.

Though Cathcart was clearly a gifted experimentalist, his major personal achievement in research lay in his extension of the dietary surveys pioneered by Noel Paton, but with an important difference. Noel Paton's attention had been directed, with reason, at the nutritional problems of the poor, especially in the great cities. Cathcart sought a wider view. Instead of looking at a single social class he studied what he called a 'vertical slice' of the town of St Andrews. To be more precise, he recruited a large sample (about one-thirteenth) of the population, chosen, so far as possible, to reflect the relative numbers of the different occupations established by the census figures. Their diets were recorded by the well-tried methods used in the previous surveys, and compared with the family income per head. As might be expected, the wealthier families had a higher intake of calories than the poor (around 3,500 calories as against 2,800). Carbohydrate intake, at around 400g per day, hardly varied with social class, but the wealthy ate more protein than the poor (100g as against 75g) and much more fat (155g as against 95g). Cathcart pointed out that this was seen in other societies: English miners, who were paid a wage 50% greater than agricultural labourers, also ate 50% more fat; and even in Bengal, where carbohydrate predominated in the diet far more than in Britain, consumption of protein, and more especially of fat, increased with increasing income. Cathcart also attempted to relate the growth of children to their diets and the social circumstances of their families. In general, he found that children in families where the foodstuffs purchased were expensive tended to be taller than in families where the foodstuffs purchased were cheap; and that the children in the (relatively few) families where 'maternal efficiency' was low were below average in height.

In the course of the same study, Cathcart addressed the essential limitation of such surveys. Measuring the consumption of the family was difficult enough. Apportioning it among father, mother and children was far more difficult.

Atwater in the United States had proposed a formula by which the requirements of an adult woman and those of children of both sexes could be expressed as fractions of those of a moderately active adult man. These had been modified by Lusk and modified again by Cathcart himself. Cathcart now compared his own scale against the actual consumption of the members of five households. The results were strikingly concordant. In each case the father received more calories than in theory he required relative to his wife and children, and this surplus was represented mainly by fat and protein. In simple language, the lion's share of whatever meat was going went to the breadwinner.

Cathcart followed up this 'vertical' study with a complementary 'horizontal' study in which he looked at the diet of the working-class in Cardiff and Reading. Their weekly expenditure was only about two-thirds of the overall average for the mixed population he had studied in St Andrews, but they consumed approximately the same amounts of protein, fat and carbohydrate. It was also similar to that of the Glasgow artisan class (a group with a relatively high income) studied by Noel Paton seven or eight years before. In every respect it was superior to that of the poor of Glasgow and Dundee. The Reading families studied were, on average, poorer than the Cardiff families, a difference reflected in a lower intake of fat. Since this was not paralleled by difference in protein intake it did not reflect a difference in consumption of meat. Again Cathcart looked for evidence of malnutrition in the conditions of the children, but found none.

While Cathcart was engaged in these studies one of his former colleagues was contemplating something on a very much larger scale. John Boyd Orr had a long, distinguished and colourful career. He was a Glasgow graduate in both Arts and Medicine. He had worked with Cathcart before the First World Was and was persuaded by him to accept the directorship of the newly established Rowett Research Institute in Aberdeen. The greater part of his research career was spent at the Rowett and his connection with Glasgow was only re-established when he was elected Rector in 1948 and Chancellor a year later. But his nutritional work marks him as very much in the Noel Paton-Cathcart intellectual tradition. Of all his contributions to nutritional science and to public life none had a greater impact than the publication in 1936 of *Food, Health and Income: Report on a survey of adequacy of Diet and Relation to Income.* In a sense this was a further development of the long series of dietary surveys conducted by Noel Paton, Cathcart and their collaborators. But it differed in one important respect. Noel Paton and Cathcart had sought to study in depth the nutrition of small groups of families. The methods they used, involving as they did the exact measurement of each family's diet for at least a week, and a careful assessment of their domestic arrangements and their management of their finances, provided an invaluable insight into the lives of the poor, but they were too laborious to be applied to the survey of large populations. Boyd Orr attempted a far less detailed investigation, but one whose results cast light on the nutritional status of the whole nation. He began by calculating the total supply of each of the major foodstuffs to the UK in 1934 and dividing it by the population to give the consumption of each item per head. He then compared these figures with corresponding figures for 1909-13 and 1924-28. The outcome was, on the face of it, reassuring: there had been virtually no change in the consumption of bread and flour, meat and potatoes. But consumption of butter, eggs, cheese, margarine, fruit and vegetables other than potatoes, all showed an increase of 33% to 88% over the 1909-13 figures. In nutritional terms, the main change was a 25% increase in the average intake of animal fats, resulting in a modest increase in energy intake from 3,057 calories a day to 3,246.

Such an improvement in terms of the population average could, of course, give no indication of how many families fell below the average and by now how much. Boyd Orr therefore collected, from a variety of sources, statistics on the food intake of 1,152 families. He classified the families into six groups on the basis of their income per head. Group I consisted of those with an income of less than ten shillings a week and corresponded to the poorest 10% of the population. Group VI, enjoying incomes of more than forty-five shillings a head, represented the wealthiest 10%. Groups II, III, IV and V enjoyed intermediate and steadily increasing incomes per head, and each represented 20% of the population. Boyd Orr then plotted graphically the estimated daily consumption of each of the major foodstuff according to his groups. The results were gratifyingly clear-cut. All groups consumed virtually identical amounts of bread, flour and potatoes, and nearly, but not quite, identical amounts of sugar. These were, of course, the cheapest foods in terms of energy. By contrast consumption, of meat, fish, eggs, fresh milk, butter, fruit and vegetables other than potatoes increased steadily as one progressed from Group I to Group VI. The differences between Group I and Group VI ranged from about threefold in the case of meat to around ninefold in the case of fruit. Clearly, intake of food high in protein (and to some extent in vitamins) depended directly on income. The consumption of only two foodstuffs fell away as one moved from Group I to Group VI: margarine and condensed milk, both cheap sources of calories. Boyd Orr had therefore established unequivocally that the nature of diet depended very sharply on income. But did it follow that this had any serious nutritional consequences? To test this he calculated the energy, the protein, the fat, the calcium, the phosphorus, the iron, the vitamin A and the vitamin C in the diets of each of his Groups, and expressed them as percentages of what were then the recommended intakes. The results again were clear-cut. In every respect the diet improved as one moved from Group I to VI and in each case the improvement was substantial - of the order of 40%. More alarming, it appeared that a substantial proportion of the population were by the accepted standards, inadequately nourished. If one looked only at the major components of the diet, energy, protein and fat, the situation was not too bad. Only Group I were deficient in all three; Group II had enough protein though fat and calories were borderline. But it was the vitamin and minerals that really gave cause for concern. If Boyd Orr's calculations were correct, not just Group I, but Groups II and III (ie, virtually half the population) had less than the recommended intake of both the vitamins and all three of the minerals included in the survey. In retrospect, the situation may not have been quite as bad as Boyd Orr believed. As his successor at the Rowett, Sir David Cuthbertson, once remarked to the writer, Boyd Orr's standards of nutritional adequacy tended to err on the high side. But his argument for a connection between nutrition and social class did not rest simply on chemical considerations. He was able to show that, age for age, public schoolboys were taller by an inch or so than boys attending a grammar school and they in turn were taller by about the same amount than boys from the working classes. The connection between diet and growth seemed to be confirmed by a celebrated experiment which he carried out with Robert Garry (another of Cathcart's pupils and ultimately his successor, in the Glasgow Chair). They compared the growth rates of two sets of rats. One of these was given a diet made up to resemble that of Group I of the population. The other received the same diet supplemented by an abundance of milk and a much green food as they could eat, thus remedying its nutritional deficiencies and making it as satisfactory as that of Group VI. The results were dramatic. The animals who received the supplement grew almost twice as fast as those on the basal diet. Boyd Orr recognised, of course, the dangers of extrapolating from rats to human populations. But in Kenya he had compared the health of the Masai and the Kikuyu, two tribes who shared the same climate and the same sort of housing but with different diets. The diet of the Masai, who were pastoralists, was high in proteins, minerals and most vitamins; the diet of the Kikuyu, who were arable farmers, was high in carbohydrate but relatively poor in most other constituents. Boyd Orr thought it approximated to that of the Groups I and II he had studied in Britain. A medical survey of the two tribes showed that Masai men were five inches taller on average than Kikuyu men; moreover the Kikuyu were subject to a wide range of diseases from which the Masai did not suffer. Boyd Orr's work is open to criticism in detail but he and Cathcart laid the intellectual foundations for the food policy followed by successive British governments from the mid-1930s through the Second World War and beyond, which was to effect so dramatic an improvement in the physique of British children.

The year 1947 marks a watershed in the history of physiology and biochemistry in Glasgow. Cathcart retired as Regius Professor and his successor in the Cathcart Chair (George Wishart) was translated to a full-time post as Dean of the Faculty of Medicine. To the Chairs thus vacated the University Court appointed respectively Robert Campbell Garry and James Norman Davidson. The two men had briefly worked together in the past when Garry had been Professor of Physiology at Dundee and Davidson his only Lecturer. As will become apparent from what follows, they were utterly different personalities and came from quite different backgrounds. For the next 25 years, until they were separated by Garry's retirement in 1970 and Davidson's sadly premature death in 1972, they were to play the commanding roles in the development of their respective departments. However much they differed from one another, they shared a common problem. The Second World War had left the University, like many other British institutions, dilapidated and impoverished. The laboratories that McKendrick had bequeathed to Noel Paton 40 years earlier were not physically well suited to modern apparatus and indeed there was little modern apparatus within them. The two new professors had only two lecturers each to cope with classes grossly swollen by demobilised ex-servicemen. Both men set about the task of rebuilding. Both, in their very different ways, succeeded brilliantly. But they did so separately. Davidson was quite clear in his own mind that physiological chemistry had outgrown the protective shell of physiology. He insisted that his Department should be independent of Garry's and that it should be rechristened Biochemistry. His staff and students regarded this emancipation favourably. Garry, while accepting it gracefully, did not conceal lingering regrets about the divorce. After all, as he plaintively but quite justifiably remarked, he regarded himself as something of a biochemist.

Davidson's appointment resulted not merely in a change of name for his Department but also in a sharp change in the direction in which it was steered. Garry, on the other hand, thought it better to redevelop Physiology along the lines he had known under Cathcart. In part, his attitudes could be traced in his background. The son of the science master at the Girls' High School, he graduated M.B.,Ch.B. with Honours in 1922, winning the Brunton Memorial Prize as the most distinguished medical graduate of his year. In spite of his academic brilliance he realised that in the harshly commercial and competitive world of contemporary medicine his lack of capital restricted his career prospects to such relatively unattractive and ill-renumerated specialties as public health and mental hospitals. He consulted Cathcart. From then on his commitment to physiology was determined. After, on Cathcart's advice, completing in the Western Infirmary the customary year as resident in medicine and surgery, Garry returned to work in Cathcart's laboratory for the next nine years. Society then was much more hierarchical then than it is now. When a boy was sent to the railway station to help Noel Paton carry parcels, Noel Paton saw to it that the boy walked, not beside him, but a respectful step or two behind. Junior doctors and junior academics stood a little higher in society, but not very much. Since the hospitals had no central biochemical services, such analyses as were performed were carried out by the resident in the side-room. Garry always took pride in having prepared well in advance for the first diabetic admitted to the Western to be treated with the newly discovered insulin. Equally, Cathcart insisted that his young men prepare their own analytical reagents, even though they could have been bought ready-made from commercial suppliers. It was, said Cathcart, good for them. There was little technical help, but it was provided by technicians of exceptional competence, one of whom went on to be a Lecturer and another to be a Church of Scotland minister. It was perhaps this early experience that instilled in Garry that concern for the proper education of technicians which marked him as professor.

Intellectually Garry benefited greatly from Cathcart's patronage and from his association with the half-dozen or so exceptionally able young people Cathcart gathered round him. There was in those days no regular system of sabbatical leave, but quite informally and without, one suspects, the cognisance of the University Court, members of Cathcart's staff would be sent off for a term or so to acquire new experience while their colleagues covered up for their absence. In this way Garry was sent in turn to Freiburg-in-Breisgau, University College London and the Physiology Department at Leeds. In 1933 he went to the Rowett Research Institute as Head of Physiology under Boyd Orr, but after two years left for the Chair of Physiology at Dundee. When he came to Glasgow in 1947 he enjoyed three important advantages. He had an established reputation in research derived from work in a broad range of fields but especially in the physiology of the caudal gut; he had 12 years of experience as Head of Department and as an academic politician in the very difficult relationship between Dundee and St Andrews; and he was coming back to the Department where he had spent his youth and whose problems and potentialities he knew intimately.

In the 23 years that he occupied the Regius Chair, Garry did much for his Department. He refurbished the by now somewhat antiquated fabric it occupied. He brought it from the era when experiments were recorded on smoked drums to the new world of electronics and the oscilloscope. He pioneered the use of diagrams rather than text in teaching. He was himself a brilliant lecturer. Even after the lapse of 45 years, his expositions of the structure of the gut and the function of the kidney are vivid in the recollection of at least one of his students. In the memoirs he wrote in his eighties he said that for him teaching was self-indulgence; and indeed at a stage in his career when self-interest would have dictated devoting all his efforts to furthering his research he involved himself in lecturing for the Workers' Educational Association and in talks for schools on what was then called the wireless.

But Garry's great achievements were essentially personal. He had a talent for identifying among the medical students who passed through his Department young men of exceptional promise. Once he had identified them, he used his great network of friendships in the Physiological Society to ensure that they could spend sabbatical years or terms in the laboratories where their professional education and their intellectual development could be advanced. In the same way he fostered the growth of the infant Department of Pharmacology as it evolved out of Materia Medica. Nor was his patronage limited to a small circle of favoured pupils. Though in his memoirs he said that the Physiology Department in his early days was to a considerable extent outside the medical establishment of the town and University, as Professor he developed excellent clinical links with the local hospitals and welcomed collaboration with surgeons.

Garry retired to Comrie in 1970, but visited Glasgow from time to time to consult the cardiologists in the Western Infirmary about his cardiac pacemaker. The journey involved four changes of buses, accepted, as one of his former lecturers remarked, because, even for Garry, it was a bit far to cycle. In the early

1980s he completed writing his memoirs, composed in the vigorous style all his students will remember, and spiced with that very occasional asperity which enlivened his lectures. In 1982 he made his last public appearance in Glasgow to present in his old lecture theatre a final demonstration to the Physiological Society on 'the physiological training of Victorian medical students in Glasgow'. He was an old man, somewhat infirm, but the focus of the immense gratitude and affection of a host of friends colleagues and former students.

In many aspects Davidson represented Garry's antithesis. In spite of his brilliant academic record, Garry always regarded himself as a young man of humble antecedents, brought up in a state school, with career prospects limited by relative poverty and with no-one to turn to for advice save the Dean of the Medical Faculty. Davidson, by contrast, was the son of the Treasurer of the Carnegie Trust and Dux of George Watson's. When advice on his choice of career was needed, his father was able to obtain it from Sir James Colquhoun Irvine, Principal of St Andrews, and one of the most eminent chemists in Britain. Garry was born and bred in Glasgow; Davidson was an Edinburgh boy whose roots lay in Buchan and Fife and who only slowly reconciled himself to his new status as a Glaswegian. Garry was primarily a medical man attracted to physiology by Cathcart, and throughout his life retained, as did Cathcart, interests centred on *Homo sapiens* and the means by which health could be improved and disease treated. At school Davidson had excelled in languages and music, but had then been attracted to chemistry, in which he gained a First at Edinburgh. He had recognised, with a prescience extraordinary in such a young man, the opportunities opening up for the application of chemistry to medical problems and took a degree in Medicine to equip himself to exploit them.

Had he wished to do so, Davidson was well placed to continue in the Cathcart tradition. He had worked with C. P. Stewart, who had established clinical biochemistry in Edinburgh and had indeed absented himself from his medical classes to do so. All his life he retained a special interest in the application of biochemistry to medicine. But his view of the science he professed was much wider. His appointment to the Gardiner Chair came at a propitious time. As he correctly discerned, biochemistry, which had been no more than the debatable land between chemistry and physiology, was becoming a discipline in its own right. The direction he gave to his, as yet small, department reflected that vision. The nutritional work which had run like a silver thread for 40 years was continued. But the new units he established represented the growing points of contemporary biochemistry: proteins, enzymes, microbial metabolism and (his own specialty) nucleic acids. He pioneered the use of radioisotopes and of tissue culture. He found money for the complex and ponderous equipment necessary for studies involving the use of electrophoresis and ultraviolet microscopy. He bombarded the University Court relentlessly with demands for more resources, more staff, more accommodation. The Principal's secretary on one occasion remembered that all his letters included somewhere the words "utterly inadequate". When a sub-committee of Court visited his laboratories he ensured that all his staff were busy at their benches doing something which generated heat or noise or, preferably, both. He succeeded to the point where one member of the sub-committee felt faint and had to retire in search of fresh air. At the same time he revolutionised the Department's teaching. He persuaded Carl Browning, the elderly and distinguished Professor of Bacteriology, to devote a considerable amount of his time to instructing Junior Honours biochemists in elementary microbiology. He similarly persuaded R. A. Robb, Lecturer in Statistics, to initiate the senior Honours class in the mysteries of standard deviation and analysis of variance.

Davidson's tactics worked. One of his lecturers, returning from a sabbatical in 1959, observed with satisfaction that the Department, though still housed in its original accommodation, was equipped to the best North American standards. Its international reputation was demonstrated by a steady flow of post-doctoral fellows. And whereas in the early 1950s barely a dozen science students were eccentric enough to study biochemistry, by 1970 something like a quarter or even a third of the students entering the Faculty were taught the subject at least to elementary level. The output of Honours graduates, which had never exceeded single figures in the 1950s, had risen to 25 by 1970, the maximum that Davidson thought the Department could teach to the standards of excellence on which he insisted. Places in the Honours school were subject to fierce competition. Davidson's career reached its zenith in the late 1960s. At long last his Department had been rehoused in new purpose-built accommodation, his academic peers had elected him F.R.S. and government had made him a C.B.E. Sadly, these successes were followed by a sequence of heart attacks to the last of which he succumbed in 1972.

Davidson was succeeded as Head of Department by Martin Smellie, whom he had recruited in 1949 only two years after his own arrival in Glasgow. Thus was established a partnership which was to endure for the rest of Davidson's life. Smellie, newly graduated from St Andrews, was the son of a Glasgow minister who had served with distinction as an officer in the First World War. After graduation he had been called up and commissioned into the Royal Scots Fusiliers. His personality reflected his background: he had the cheerful ebullience so often seen in sons of the manse; and he had the drive, initiative and doggedness in the face of difficulties that the British army expects of its subalterns. He was to find great scope for these qualities in his new appointment. For biochemists, especially in Britain, the late 1940s and early 1950s were a period of both excitement and frustration. Excitement because it was becoming clear (as Davidson had previously discerned) that biochemistry was capable of shedding light on a far wider and deeper range of phenomena than had hitherto been realised. Frustration because the methods and equipment available to bring this about were so limited and unreliable and yielded so little information in relation to the effort expended. In the 1990s, when laboratories are full of elegant, automated, reliable electronic equipment, it takes a considerable effort to recall how difficult things were 40 years ago when electronics depended on the thermionic value, not the transistor; when the range of commercially available reagents was very limited; when the only way to procure an enzyme was to learn how to purify it for oneself; when much apparatus was home-made; when experimental work was always in danger of being interrupted by power cuts.

Smellie positively enjoyed the challenges of this Spartan environment. He began to achieve a public reputation by exploiting an electrophoresis apparatus of his own devising, set up from army surplus equipment and well-calculated to electrocute an incautious operator or to set itself on fire. (Happily it did neither, though it did smoulder from time to time.) He seemed able to turn his hand to anything. When the supply of so simple a reagent as hydrogen peroxide failed, he set out to manufacture his own. When a fraction-collector or spectrophotometer failed to function, he could generally find the fault and put it right. When carbon-14 joined phosphorus-32 as an isotopic tracer it was he who put into commission the Heath-Robinson equipment necessary to count its low-energy β-particles. These efforts brought their reward. He was commissioned to write his first major review before he even graduated Ph.D. and shortly thereafter he became effectively day-to-day director of Davidson's own research group. His success in that role and his discharge of the many departmental duties thrust upon him resulted in rapid promotion, and when in the mid-1960s a second Chair of Biochemistry was established and named after Cathcart it was natural that he should be appointed to it. It was equally natural that on Davidson's death he should be appointed his successor.

The problems Smellie faced in 1972 were different from those that had confronted Davidson 25 years earlier but they were no less daunting. Davidson's term of office had coincided with that long post-war period in which the British universities were held in high public esteem and successive governments were in general prepared to fund them ever more generously in the hope that the money so spent would prove an investment that would yield a handsome return in the not too distant future. This happy state of affairs in some degree made Davidson's objectives less difficult to attain than might otherwise have been the case. By 1972 the tide had began to turn. The hoped-for 'white-hot technological revolution' had not manifested itself. The insolence and intolerance of a handful of student radicals, sometimes incited by their elders, alienated the public and tarnished the reputation of the universities as a whole. The values and ideals which had guided them for centuries began to be questioned. Public and politicians alike demanded assurance that the money they had so generously provided was being applied effectively to the ends they sought. The tide of popular support began to ebb. It is ebbing still.

It was in these unpropitious circumstances that Smellie took over. There is no sign that he was daunted by them any more then he had been by the dearth of resource when he was a young Assistant. His response was cautiously and carefully to diversify the Department's research activities. He was a man of wide scientific interests and shrewd scientific judgment with extensive contacts through the Biochemical Society and the Royal Society of Edinburgh, in both of which he played an active role. These allowed him to recruit judiciously. His policy of diversification was given impetus by the appointment to the Gardiner Chair first of Alan Williamson (1973-81) and then of Miles Houslay (1986 onward). Both were young men of proven distinction in research. Both introduced new blood and new interests, immunology in the case of Williamson, molecular pharmacology in the case of Houslay.

Sadly, like Davidson, Smellie was not spared to enjoy the fruits of all he had laboured to effect. He had been appointed Vice Principal, an office for which his great administrative talents and warmth of personality admirably fitted him. He was taken ill at a committee meeting one Friday morning and admitted to the Western Infirmary, for observation rather than anything else. Though he had a history of coronary artery disease, his friends were not unduly worried: a few days enforced rest would do him good. But in the small hours of the following morning his condition suddenly deteriorated and he died. His colleagues had been well aware of his medical history and recognised that his health was precarious. But that he should die just short of his 62nd birthday shocked them. He was not merely senior professor. He had been a central figure in the development of the Department for almost 40 years. His loss was deeply mourned.

It used to be the fashion for those who wrote textbooks of modern history to select a terminus when history ceased and current affairs began. Smellie's death might be taken as such a point. But a coda may be added. The Department he had done so much to build survived his passing. Indeed under Houslay's leadership its reputation has steadily risen. Smellie would have been proud of it. And so would Davidson, and indeed Cathcart.

As Smellie succeeded to the Headship of Biochemistry on Davidson's death so two years previously, Ian Alexander Boyd had succeeded Garry in Physiology. Superficially Boyd and Smellie had a good deal in common. They were close contemporaries. Both had their roots in Glasgow's professional middle class. Both had served a long apprenticeship in their respective departments and were regarded as the obvious candidates to take over from their predecessors. Both had been appointed in the 1960s as the first incumbents of newly-created second Chairs in their respective departments. But there the similarities ended. Smellie was, not merely physically, a figure larger than life. Gifted with a commanding personality, he was superbly equipped for what his predecessor had defined as the principal role of a Scots professor: the creation of a department in which younger men and women could do good work. The same qualities made him a prominent figure in university affairs. Boyd was by comparison a quiet, rather reserved man, whose real qualities were less immediately apparent outside his own circle. He has been described quite simply as the most distinguished Scottish physiologist of his generation. His experiments on muscle spindles (on which he was a world authority) were notable for their elegance and for the manner in which every last particle of significance was extracted from the results they yielded. His microcinematographic films have been used world-wide in the teaching of physiology. His work was demanding physically and mentally - experiments might stretch without intermission for 24 hours - but his dedication commanded the affection as well as the respect of his collaborators. He was equally successful as a teacher, a lucid and stimulating lecturer with an air of authority that amounted to charisma. He did not relish administration or the business of university government (whereas Smellie did), but he brought to it that high conscientiousness and meticulous attention that characterised all he undertook. As Head of Department he showed a scrupulous fairness and kindness, especially to young people setting out on careers in research. His international reputation made it relatively easy to attract outside funds at a time when such support was becoming increasingly important, and in a period of rapid technological change fostered electron microscopy as one of the Department's specialities. He was equally effective in his work for the Faculty of Medicine. Indeed a colleague once described him as "our preclinical dean".

In 1983, having served his department as Professor for over 13 years, first of expansion then of contracting resources, Boyd, then aged only 56, took formal premature retirement from the service of the University, though he continued, in effect, with full-time teaching and research. And he continued his life-long work for the Scottish Schoolboys' Club, which his father had founded. But he was not long to enjoy freedom to pursue the work for which he felt himself best fitted. He died suddenly while on a brief holiday in Wales in the autumn of 1987. It is said that in noting the fact, the *Glasgow Herald* referred simply to his prominence as a youth leader.

Garry's successor as Regius Professor was Otto Hutter. His appointment marked a break with tradition. All previous incumbents of the Chair had been Scotsmen, all had been physicians and graduates of Scottish universities. Hutter was an Austrian refugee, one of that intellectual diaspora which has so enriched science and scholarship in Britain, as in the United States, for the past 60 years. He was not a graduate in medicine. His first employment was with Burroughs Wellcome and his first degree was obtained at University College London. But his reputation was soon established. His work with Trautwein on the pacemaker activity of the heart is such a classic that the tracings from their experiments are commonly reproduced in undergraduate textbooks. By the time Hutter came to Glasgow his authority in electrophysiology and in neuromuscular transmission was generally acknowledged. While at Glasgow he worked chiefly on the mechanism of potassium ion penetration and of potassium ion conductance in skeletal muscle. Like Boyd, he proved a gifted teacher. No one who heard his lecture to a lay audience on 'Harvey and the Circulation of the Blood' could fail to be deeply impressed by his sense of history, his power of exposition, and his feelings for language. But his forte was not so much the set-piece oration but the tutorial with a small group of students, to whom he communicated his sense of intellectual excitement. One participant in such an occasion, himself the son of a Glasgow lecturer and an unsparing critic of his father's colleagues, once remarked: "We were supposed to have Otto for an hour, but he got so enthusiastic he went on for two. You can't really help admiring somebody as keen as that". Apart from his personal teaching, one of Hutter's main achievements for his Department was the re-equipping of the junior teaching laboratories with modern electrophysiological equipment. In the outside world he played a major role through the International Union for Physiological Science, in raising the standards of undergraduate education worldwide. In that capacity he visited Shanghai as a member of a delegation to introduce modern teaching of electrophysiology into China.

Hutter retired from the Headship of the Department in 1990 and from the Regius Chair years later. He has latterly developed a totally new interest in membrane fragility in muscular dystrophy, which he still pursues when he can spare time from gardening. His more orthodox colleagues suspect that he may eventually turn them into a molecular biologist.

As Smellie's tragically premature death may be taken as a terminus in the history of the Biochemistry Department, so may Hutter's retirement in the case of Physiology. Again there is a happy coda. Just as Biochemistry has thrived in recent years under Miles Houslay, so has Physiology under, successively, Sheila Jennett and Ian McGrath. Cathcart, were he alive, would hardly understand the search in which it is engaged or even the teaching it offers. Garry, if age and infirmity had permitted him to visit his old Department, would have had no difficulty in understanding what was going on, but might wonder at the sophistication of the apparatus available to his successors. But both would be pleased by what they saw.

The history of anatomy in Glasgow is quite different from that of physiology and biochemistry. In part this reflects the nature of the three subjects. Anatomy is by far the oldest of the sciences underlying medicine. The structure of the human body in gross terms of bone and muscle, nerves and viscera could be explored and described by anyone gifted with a sharp scalpel, keen eyesight, a talent for drawing, and a source of cadavers. Physiology, and still more biochemistry, had to await the invention of the physical and chemical techniques by which they could be advanced. The structure of the heart could be, and was, accurately described long before Harvey demonstrated that the blood circulates rather than merely ebbing and flowing. It is not therefore surprising that a Chair of Anatomy (albeit conjoined with Botany) should have been established in 1718 (making it one of the earliest in the University's history) while Physiology had to wait till 1839 and Biochemistry till 1919.

Departments are a fairly recent feature of university life, but the present Anatomy Department may look for its origin to 1720, when Thomas Brisbane was appointed to the newly created Chair of Anatomy. Though Brisbane and his immediate successors were all medically qualified, their interests were wide ranging, and included from the first what would be recognised as 'science'. Indeed the Chair was associated formally with Botany until 1818, and Brisbane himself was by inclination a botanist. Later, in 1756, the Chair was held briefly by the distinguished chemist Dr Joseph Black. Some of the early anatomists also had the scientist's fondness for practical work, and arrangements for practical anatomy were being discussed as early as 1741. Anatomy, from vocational necessity, may indeed have set the pace for practical work in which students actually participated. By the 1820s the Glasgow anatomist Granville Sharp Pattison could claim that a surgeon should have dissected 12 subjects during his medical education, whereas modern practical courses in natural philosophy were not started by Lord Kelvin till the 1850s.

Anatomy can also lay claim to one of the great pioneers in scientific method, in William Hunter who attended classes in the University from 1731 to 1736. Avidly dissecting, drawing and writing, Hunter was one of the rugged empiricists of his time, able to grasp general principles but equally able to apply his results to the immediate improvement of his practice.

Formal connections of anatomy with modern science in the University can be traced to 1872, when the new degree of B.Sc. was introduced. The requirements for this degree included four subjects from the Arts curriculum, and four subjects from a specified list of scientific subjects. From the first, the science options included anatomy and other medical classes and this gave the emerging Department a smooth and natural entry into the Faculty of Science in 1893.

When the Science Faculty was becoming established, Anatomy was settled into its new purpose built building, which extended to the east beyond the University. This building consisted of a classroom and sideroom, a dissecting room and mortuary, and a demonstration gallery. These rooms are still recognisable, though much altered, as the present anatomy classroom, the dental dissecting suite and the southernmost bay of the museum. At this time the building was entered by a staircase which led down from the range of classrooms closing the eastern side of the East (or Science) Quadrangle. These classrooms housed the medical departments of the time, including Natural History. In 1907 many of these moved out to the new West Medical Building. Others retained their classrooms here until the 1960s, but all have now gone, leaving Anatomy as the sole occupant of the original site, and rather awkwardly divorced from its natural partners.

The Professor of Anatomy in 1892 was John Cleland, whose 'great leonine head' was the subject of a magnificent contemporary bronze cast. Cleland, described in his obituary as the 'Nestor of British Anatomy' had been Demonstrator to Goodsir in Edinburgh, and a colleague of Sir William Turner, a notable 19th century anatomist, who became President of the General Medical Council. In 1892 Turner was also an editor of the *Journal of Anatomy and Physiology*, with his colleague J. G. McKendrick, Professor of Physiology in Glasgow. Anatomy at this time was deeply influenced by Darwin, and anatomical research was concerned frequently with comparative studies, structural variations and palaeontology. Cleland's early research was in these fields, and would now be of little interest, though in 1878 he published an article 'on the cutaneous ligaments of the phalanges' - ligaments which have lately become unexpectedly important in the reconstruction of the finger in modern plastic surgery. The later part of Cleland's career was dominated by technical advances in histology, physiology and embryology. The modern microtome for cutting histological sections appeared in 1870, and formalin fixation was introduced by Blum in the 1890s. It was during this time that histology is first mentioned in the teaching programme of the Department, and a Lecturer in Embryology was appointed in 1894.

In 1901 through the generosity of the executors of James Brown Thomson, a handsome extension was built on the north side of the existing building. This extension was designed by Sir John Burnet, in an entertaining fin-de-siecle version of Scots renaissance, complete with corbie steps and bartizans. Inside there are some original fittings and some muted Art Nouveau. The new building contained staffrooms, an anthropological laboratory and the splendid museum hall. This now holds the anatomical part of the Hunterian Collection, one of the most important anatomical collections in the country. It is displayed attractively in modern cases, but retains its original hand-made 18th century bottles. The collection is essentially a research collection, and formed the basis of Hunter's publications on the functional anatomy of the lymphatic system and the gravid uterus. The hall also houses the collections of other professors of Anatomy, namely Hamilton, Cleland, Jeffrey and Allen Thomson. Thomson, Professor from 1848 to 1877 was convener of the committee which removed the University from High Street, and advised on the design of the original science classrooms at Gilmorehill.

The leading figure in Anatomy in Glasgow in the early part of the 20th century was Thomas H. Bryce, who was appointed to the Chair in 1909. Bryce came of a distinguished Scots family, and was a cousin of the historian and statesman Viscount Bryce. Professor Bryce became a legend to a generation of undergraduates, who are only now passing away, and was the subject of a legion of possibly apocryphal stories and rhymes. He even figures in the memoirs of James Bridie, who was one of his students. As an anatomist, Bryce was fairly conservative, and was essentially a topographer and anthropologist in the tradition of Thomson and Cleland. However, within these fields he was highly active. In 1896 he communicated a paper on movements of his own wrist as seen by X-ray. This paper, written in co-operation with Glasgow's pioneer radiologist Dr John Macintyre, is probably one of the very first papers on normal radiological anatomy. Bryce also edited *Quain* (a powerful rival of *Gray's Anatomy*), with Sir E. Sharpey Scafer and Johnston Symington, giants of British histology and anatomy. Bryce's contribution may well be seen as the high point of a species of evolutionarily inspired topographical anatomy, which has now fallen entirely from favour, but which remains a classic of its kind. Bryce is also remembered for his association with the Bryce-Teacher embryo, a very early and puzzling human embryo, which in the light of later investigations seems to have undergone pathological change.

With the development of Bryce, and the appointment in quick succession of two of his pupils, D. MacCallum Blair (1935-42) and William Hamilton (1942-46) there came a sharp swing away from traditional topographical and anthropological concerns, and the rapid rise of new science-based teaching and research. In teaching, modern histology and neurohistology began to displace traditional topographical anatomy, which has seen a sharp reduction of hours through the 1950s and '60s to the present time. In 1934 Blair, whose innovations were widely copied in the UK, foresaw a united Department of preclinical science with specialists in embryology, topography, neuroanatomy, metabolism and biochemistry - a vision uncannily close to some of the current proposals for preclinical integration. In research, the experimental approach to biological problems, so reminiscent of Hunter, began to revive. Under Hamilton, promising lines involved the living mammalian ovum, hormones in development and the immunology of implantation. At the same time, Blair's initiatives in neuroanatomy were greatly stimulated by war work on nerve grafts by Wyburn and Bacsich.

These lively developments continued into the '50s and '60s under the quirky and imaginative leadership of G. M. Wyburn. In 1943 Tom Gibson and Peter Medawar published a clinical research paper entitled *The fate of skin homografts in man,* based on work carried out by Gibson in the Royal Infirmary in Glasgow. Their work stimulated a fruitful research collaboration between Gibson's colleagues in Plastic Surgery and the Department of Anatomy, and in the 1950s Tough, McGregor and R. J. Scothorne were beginning to unravel the basic immunology of graft rejection. Wyburn had the foresight to see the importance of electron microscopy, and with the help of Dawson of Chemistry, one of the fathers of this technique, he established the first biological electron microscope unit in the University. The first microscope was a Philips EM75b, and was in almost constant use night and day. Wyburn himself was involved in a mammoth electron microscope study of the structure of the oviduct of the hen, in relation to egg shell formation - a thoroughly modern functional investigation with commercial overtones, undertaken in association with the Veterinary College. A. D. Hally applied electron microscopy to the problem of acid secretion in the stomach, and won the Bellahouston Medal of the University for his results. The early '60s were an exciting time, when Wyburn gathered round him a cadre of young doctors and scientists - Calman (now the Government's Chief Medical Officer) and Baillie in histochemistry, Scothorne and Warbrick in embryology, Clarke in X-ray microscopy, and Hally and Toner in electron microscopy. Under Scothorne the Department was extended to accommodate these developments in teaching and research. New classrooms equipped with closed circuit television were opened, new staff engaged, and a new tissue culture research laboratory opened, this last named felicitously after Allen Thomson.

Today the Department is still based on the original nucleus of classroom, dissecting room and museum; but it has extended from this in all directions. In the museum, immunocytochemistry, CT scans and computer reconstructions jostle with the Hunterian lymphatics. Teaching has expanded to cover a wide range of new undergraduates and professional classes, and the focus of research has shifted towards neuroscience, which is nationally one of the major areas of current anatomical development, and a powerful research initiative within the University

These recent developments have subtly changed relationship of Anatomy to other preclinical subjects and to the Faculty of Science. In 1720, Anatomy was the only preclinical science. Today, it is one of many, some of which by accident of history or discovery have surged forward and overtaken it. At the same time the relationship with the Faculty of Science has been strengthened. It seems likely that the option of anatomy as a science subject in the B.Sc. took the Department of Anatomy into the Faculty of Science in 1892. In the 1930s the

B.Sc. had replaced the M.A. as a preliminary degree for many entering medicine. After the War, the Honours B.Sc. became a popular degree with a select group of well qualified medical students. The anatomy option offered the opportunity to take part in genuine original research, which almost universally led to scientific papers and publications and often to visits to overseas laboratories. Currently, there is a vigorous development of a new science school of undergraduates who are studying anatomy as a biological science, without any direct medical connection. This trend is also reflected in the composition of the teaching staff, who are now all qualified in science, with only a small number holding medical degrees as well. Anatomy is emerging as an independent biological science in its own right. It has justified its original decision to keep a place among the other biological sciences in the Faculty of Science.

Acknowledgements
We are deeply grateful for the help of Dr Margaret H. Gladden and Dr David J. Miller. Any errors are, of course, our own.

Robert Young Thomson is Grieve Lecturer of Biochemistry and John Shaw Dunn is a Senior Lecturer in the Department of Anatomy at the University of Glasgow.